Biology's First Law

Biology's First Law
The Tendency for Diversity and Complexity
to Increase in Evolutionary Systems

Daniel W. McShea and Robert N. Brandon

The University of Chicago Press :: Chicago and London

Daniel W. McShea is associate professor of biology, with a secondary appointment in philosophy, and Robert N. Brandon is professor of philosophy, with a secondary appointment in biology, both at Duke University.

The University of Chicago Press, Chicago 60637
The University of Chicago Press, Ltd., London
© 2010 by Daniel W. McShea and Robert N. Brandon
All rights reserved. Published 2010
Printed in the United States of America

18 17 16 15 14 13 12 11 10   2 3 4 5

ISBN-13: 978-0-226-56225-4 (cloth)
ISBN-13: 978-0-226-56226-1 (paper)

ISBN-10: 0-226-56225-5 (cloth)
ISBN-10: 0-226-56226-3 (paper)

Library of Congress Cataloging-in-Publication Data

McShea, Daniel W.
   Biology's first law : the tendency for diversity and complexity to increase in evolutionary systems / Daniel W. McShea and Robert N. Brandon.
       p. cm.
   Includes bibliographical references and index.
   ISBN-13: 978-0-226-56225-4 (cloth : alk. paper)
   ISBN-13: 978-0-226-56226-1 (pbk. : alk. paper)
   ISBN-10: 0-226-56225-5 (cloth : alk. paper)
   ISBN-10: 0-226-56226-3 (pbk. : alk. paper)   1. Evolution (Biology)
2. Evolution (Biology)—Philosophy. 3. Biodiversity. 4. Biocomplexity.
I. Brandon, Robert N. II. Title
   QH360.5.M38 2010
   576.8—dc22

                                                        2009044828

For Diane, Hannah, and David

For Gloria and Katherine

Contents

Preface

So certain is this that we may boldly state that it is absurd for human beings even to attempt it, or to hope that perhaps some day another Newton might arise who would explain to us, in terms of natural laws unordered by intention, how even a mere blade of grass is produced. **Kant,** *Critique of Judgment* **(1790)**

The book in your hands shows Kant to be wrong. It articulates and defends the zero-force evolutionary law, evolution's analog to Newton's first law, the law of inertia.

We did not set out to write a book refuting Kant. Kant has been shown wrong on so many points that such a task would not have motivated us. Nor do we wish to compare ourselves to the incomparable Newton, for we fear that there is some chance that we might come up on the short end of that comparison. Finally, neither of us suffers from physics envy (at least according to our friends and families). So what explains this book?

Here's the story. The ideas behind this work have been fermenting in the heads of the coauthors for decades. One of us (Brandon) is a philosopher, who argued as early as 1978 that to understand natural selection one had to be able to distinguish differential reproduction due to selection from that due to chance. Accordingly, no understanding of selection is possible without a conceptually clear, and operationalizable, conception of drift. Thus, from an interest in natural selection Brandon was led to an interest in drift. Many of his publications (Brandon and Carson 1996; Brandon and Nijhout 2006; Brandon 2005, 2006) are intellectual ancestors to the present work. The other of us (McShea) is a paleobiologist who has written about complexity in evolution for 20 years now. In a 1991 review, he briefly discussed Herbert Spencer's work on evolution, in particular Spencer's principle of "the instability of the homogeneous" (Spencer 1900). Given a homogeneous object, different accidents will happen to different parts of it, producing a heterogeneous object, or, in the terms we develop in this book, a more complex one. The idea lay dormant for some years, reemerging again in a 2005 paper for a volume in honor of Stephen Jay Gould (McShea 2005a). It was Gould, after all, who informed our collective consciousness in evolutionary studies about the central importance of chance in macroevolution. How better to honor him than to return to Spencer, an entire metaphysic with accidents, chance, at its heart. Finally, after Spencer and Gould, a third intellectual ancestor is David M. Raup. As will be obvious to those who know Raup's work, his thinking on the dynamics of randomly evolving clades is foundational, both for the 2005 paper and for this book.

So we each have our own intellectual precedents, but the present book is collaborative. And we will get to the origin of this collaboration, but first some deep background. We have known each other since 1978, when we were both *very* young men, getting degrees at the same university. (One of us was a graduate student and an undergraduate-thesis mentor and reader for the other.) We went our separate ways after that but ended up together again—improbably—at Duke in 1996. Here we coteach courses, and we are both participants in Duke's biweekly Philosophy of Biology Discussion Group. On a personal level, we tolerate each other about as well as could be expected from a pair consisting of one diehard Red Sox fan and one who bleeds Yankee pinstripe blue (a neat feat that). We will not dwell on how these allegiances came about. But we will note, as Mark Twain (1963) did, that people don't pick their religion. They inherit it from their parents. Similarly for baseball. Imprinting on a baseball team is a product of the contingencies of time and place of one's upbringing. That is how a working-class populist can root for the most corporate, buttoned-down team in professional sports. And that also explains how a devout agnostic about almost everything could for 35 years believe in a team that, during that stretch, won not a single championship.[1]

So we have known each other for a long time. But the precise date (precise enough for a paleobiologist) of the start of this collaboration is quite recent. It can be traced to Brandon's handing a manuscript, the one that ultimately would be published in the *Journal of Philosophy* in 2006, to McShea in the fall of 2005. McShea read the manuscript and found himself unusually enthusiastic. And he discovered that his enthusiasm arose from his having already made what seemed to him the same argument (McShea 2005a, 2005b). But was it really the same? Brandon's work was on drifting means, while McShea's was on diffusing variances. Brandon's was microevolutionarily oriented; McShea's was macro. McShea was talking about a force or drive, while Brandon was insisting that what he was thinking about should be thought of as a zero-force condition. Was it really the same idea? It seemed obvious that it was. After four or five months of hard thought and back-and-forth argument, we concluded that the arguments were indeed the same. And further we concluded that, yes, in fact, it was obvious. Thus, the birth of the book.

Well, not immediately. First there was the idea of writing a short article that summarized the main points and their implications for diversity and complexity. That didn't fly. This book tries to effect a fundamental gestalt shift in how we view evolutionary phenomena. It is built, not on surprising new discoveries, but on a variety of things already well known among evolutionary biologists in scattered specialties ranging from molecular evolution to paleobiology. How all of this adds up to a fundamentally new view of evolution is hard to convey. We hope to have come close to doing so in this book. We had no hope of doing so in a short article.

A word on hubris. It may seem to some that by claiming to have found a law of evolution we have invited the wrath of one or more Greek gods, perhaps the wrath of Athena, the goddess of wisdom—and if not of gods, then at least of certain contemporary colleagues for whom one of the essences of Darwinism is its recognition of the particularity of biology, the uniqueness of every individual, every species, every ecological circumstance. From this follows, some would say, an unpredictability, an absence of regularity, that must be simply accepted if not celebrated. But there is another view, not in ascendance in our time but with roots in Darwin as well, that holds out a hope of finding regularity. It has always been just a hope. One cannot be a Darwinian without admitting the possibility that history might be just, as Henry Ford put it, one damn thing after another. But, this alternative view asks, wouldn't it be exciting, wouldn't biology be so much more fun and fascinating, if laws did exist, if amid the blooming buzzing confusion of evolutionary change there were some underlying order? Now we do not claim to have found a first-order regularity. We do not claim to be able to predict how any given species will change. What we think we have is a second-order regularity, one that governs not the precise changes occurring in evolution but the distributions of those changes. Further, and luckily for us, we can advance this more modest regularity while claiming little originality for its discovery. As we explain later, the underlying principle is present, in various disguises, in subfields throughout biology. So we think, we hope, that we are safe from hubris. We think Athena would approve.

And indeed, when all is said and done, we think Kant would approve as well. Kant was an exponent of explanation by natural law. His skepticism about laws in biology came from his doubts about our ever explaining the teleological aspects of biology in terms of laws. Kant, like most of his contemporaries, saw nature as being thoroughly infused with purpose. Of course, neo-Darwinians find this less problematic than did Kant, because they think that natural selection can explain apparent teleology (i.e., adaptation) in a scientifically respectable way. We agree with that. But we think there is more to evolution than adaptation. There is diversity and complexity. And there is no reason that there could not be laws about these. In particular, we think—and we hope to have shown in this book—that there is always and everywhere a tendency for diversity and complexity to increase, a tendency that does not depend on natural selection. We boldly state that Kant would have been pleased.

Acknowledgments

Steve Gould, Dave Raup, Stan Salthe, Leigh Van Valen, and Bill Wimsatt are all implicated in various ways in this work. My gratitude to all of them and apologies to those who would rather not have been. Thanks also to those who read all or part of the manuscript, as well as to the many who discussed the ideas in this book with me and provided the support on which any undertaking such as this depends: Carl Anderson, Kristen Ban, Gabe Byars, Viviane Callier, Chuck Ciampaglio, Tony Dajer, David Garfield, Michael Gilroy, Bill Goldberg, Benedikt Hallgrimsson, Rob Kunzig, Anne Lougée, Paul Manos, Jon Marcot, Lauren McCall, Dave McCandlish, Peter McIsaac, Naomi McShea, Robert McShea, Sarah McShea, John Mercer, Bill Morris, Dave Raup, Rebecca Sealfon, Kriti Sharma, Carl Simpson, Jenny Tung, Ed Venit, and Will Wilson. To this list must be added a number of people who provided expert advice on particular issues: Teddy Gray, Arnie Miller, Phil Novack-Gottshall, Mohamed Noor, Todd Oakley, and Greg Wray. I am indebted to all of you. And thanks to Greg Lamm, in the delightful ambience of whose coffee shop half of this book was written. Finally, thanks, Robert, for your clarity and precision of thought throughout this project, and for refraining from beating me with a stick as a solution to my frequent obtuseness. [DM]

First, I want to thank Richard Lewontin. Dick read through an early draft of this book and offered a number of comments that made their way into the final version. He remains a constant source of stimulation. Francesca Merlin gave me very helpful comments on an early presentation of some of the ideas contained in this book. Also, Mark Rausher and Fred Nijhout helped me clarify some ideas about drift that were precursors to the present work. My thanks to all of them. Finally, I want to thank my friend Janis Antonovics, who showed me that rigor, creativity, and fun are not mutually exclusive when thinking about evolution. Oh, and thanks Dan for making me do this. [RB]

We both extend our thanks to the students and faculty of a biweekly Philosophy of Biology Discussion Group here at Duke University, especially our closest colleagues, Alex Rosenberg, Fred Nijhout, and Louise Roth. Valuable input also came from the students in our spring 2009 graduate course: David Crawford, Paul Durst, Chris Iacoboni, Eve Marion, Carlos Mariscal, Heather Mayer, and Leonore Miller. Finally, we are grateful for the expert advice of our editor, Christie Henry, of the University of Chicago Press. Carlos Mariscal composed the elegant drawing of the picket fence in chapter 1 and helped design the cover. Heather Mayer and David Crawford kindly drafted the index for us. And Pam Bruton was our alert copy editor. Lastly, we are grateful for the honest and constructive criticism of several anonymous reviewers, who forced us to think, and then rethink, critical aspects of how the ideas in this book were presented. [DM and RB]

1

The Zero-Force Evolutionary Law

The history of life presents three great sources of wonder. One is adaptation, the marvelous fit between organism and environment. The other two are diversity and complexity, the huge variety of living forms today and the enormous complexity of their internal structure. Natural selection explains adaptation. But what explains diversity and complexity?

Evolutionary theory offers a number of possibilities. Diversity might be explained by natural selection favoring differentiation of closely related species, perhaps for the avoidance of competition (Darwin [1859] 1964). Or it could be the result of selection favoring groups of species with a greater propensity for speciation (Stanley 1979). The complexity of organisms could be favored on account of the selective advantages of greater division of labor among their parts (Darwin [1859] 1964). Or it could be that greater complexity means greater ecological specialization, which in turn might be generally favored by selection. Diversity and complexity could also be mutually reinforcing. As species diversity increases, niches become more complex (because niches are partly defined by existing species). The more complex niches are then filled by more complex organisms, which further increases niche complexity, and so on (Waddington 1969). There are many other possibilities, most relying on natural selection.

While there is undoubtedly much truth in these hypotheses, we argue that they are incomplete, that there exists in evolution a spontaneous tendency for diversity and complexity to increase, one that acts whether natural selection is present or not. To put it another way, rising diversity and complexity are the null expectation, the predicted outcome for evolution in the total absence of selection and other forces. They are the zero-force expectation. The reason is simply that variation arises in biological systems, and when heritable it accumulates, with the result that variances tend to increase. And both diversity and complexity are aspects of variance. Diversity in a general sense is a function of the amount of variation among individuals. And in the absence of constraints or forces, the accumulation of variation in a population will tend to increase the diversity of its component individuals. (More narrowly, diversity is number of species, of course, but speciation is just a special case in which variation becomes discontinuous.) Complexity in a general sense is a compound notion connoting an uncertain mix of multiplicity of parts, adaptation, functional sophistication, and more. Here, however, we adopt a narrower, technical understanding from biology, in which complexity is a function only of the amount of differentiation among parts within an individual. (Or in the special case in which variation is discontinuous, complexity is number of part types.) Again, in the absence of constraints or forces, the accumulation of variation in the parts of individuals in a lineage will tend to raise the variance among those parts and therefore increase the complexity of the individual.[1]

The rationale is simple. Imagine a new picket fence, in which each picket is identical to every other. With the passage of time, different accidents happen to different pickets. A pollen grain stains one. A passing animal knocks a chip of paint off another. The bottom of a third picket becomes moldy and crumbles where it touches the ground. As a result, the pickets become more different from each other. And the process continues indefinitely, so that even when the pickets are quite differentiated, further accumulation of accidents will tend to make them more different yet. In other words, the fence as a whole becomes more "complex," coming to consist of parts—pickets—that are ever more different from each other. Or from another perspective we might say the pickets become ever more "diverse." No directed forces need to be invoked here. The cause of the complexification of the fence (or equivalently the diversification of the pickets) is not gravity, electromagnetism, natural selection, or any other natural force acting on its complexity. Nor is any directed human intervention required. Rather, diversity and complexity

FIGURE 1

arise by the simple accumulation of accidents, producing a steady, background increasing tendency. In other words, increasing complexity and diversity is the natural state of the system, the expectation in the total absence of any forces acting on diversity or complexity.

The claim is that this same increasing tendency is the natural or background condition of evolving populations and organisms. We call the principle that governs this background condition the "zero-force evolutionary law," or ZFEL. Here is one formulation of the law:

> ZFEL (special formulation): In any evolutionary system in which there is variation and heredity, in the absence of natural selection, other forces, and constraints acting on diversity or complexity, diversity and complexity will increase on average.

The absence of constraints and forces is the special zero-force condition after which the law is named. What are these constraints and forces? Regarding forces, natural selection is far and away the best known, but we also need to accommodate the possibility that other forces may exist[2] or may someday be discovered. The special formulation of the ZFEL excludes all forces. Generally speaking, "constraints" refers to what are commonly called developmental constraints, limitations on change arising from the organization of ontogeny, as well as to historical constraints, limitations arising in evolution that are generally manifest in development. It also refers to physical and material constraints, limitations arising from the laws of nature and the properties of materials, as well as logical and mathematical constraints. In chapter 2, we state more precisely the sense in which these are excluded in the special formulation.

Notice that the special formulation does not say that diversity and complexity *will* in fact increase in every instance, only that they will increase on average. The reason is that chance could intervene. In other words, the increase is probabilistic, so that improbable combinations of events could cause diversity and complexity to fail to increase, or even to decrease. After diverging initially, the pickets in the fence could by chance become more similar to each other.

There is also a second formulation, a more general statement of the law, which does not invoke the zero-force condition:

> ZFEL (general formulation): In any evolutionary system in which there is variation and heredity, there is a tendency for diversity and complexity to increase, one that is always present but may be opposed or augmented by natural selection, other forces, or constraints acting on diversity or complexity.

What the general formulation tells us first of all is that the cause of increase in the special formulation is a tendency. It is this tendency that manifests itself as increase, on average, whenever constraints and forces are absent. Second, it says that this tendency is to be understood as present even when constraints and forces do act. For example, the tendency could be blocked by constraints, or opposed and even overcome by selection, with the result that diversity or complexity does not in fact increase, even on average. Or selection could oppose the tendency strongly enough to overwhelm it, so that diversity or complexity actually decreases. The general formulation says that, when these contrary forces or constraints act, it is not because the ZFEL tendency has vanished. Rather, the tendency remains and continues to act even while constraints or forces are overcoming or overwhelming it. Analogously, the picket fence may not actually become more complex, if someone is regularly fixing and painting it. But the complexification tendency is understood to be present and acting anyway, continuously, even while the repair process is going on.

Finally, the general formulation allows that forces or constraints may also favor diversity or complexity, augmenting the ZFEL tendency. Diversity or complexity increases more than it would if either the ZFEL tendency or the augmenting constraint or force were absent. In sum, the ZFEL tendency is to be understood as a background state that is present prior to and during the imposition of any constraints or forces. In this view, the effect of any imposed constraint or force is always the resultant of two factors, one of which is the ZFEL.[3]

A Unification

We do not consider the ZFEL a new discovery. All it really says is that there is a tendency for variation to arise and to accumulate, and that it will do so unless opposed in some way. And some such principle has been part of the implicit working knowledge of every evolutionist since Darwin. Further, many applications of the principle have long been known and appreciated in biology. For example, it is known that, in the absence of selection, a population tends to diversify as a result of mutation, recombination, and random mating (Gould and Lewontin 1979). At the molecular level, a spontaneous diversifying tendency is implicit in the methods used to isolate the effect of selection from drift (Kreitman 2000; Yang and Bielawski 2000; Bamshad and Wooding 2003). And the ZFEL is clearly central in recent work on gene duplication and divergence, especially the work of Lynch (2007b; also Ohno 1970; Taylor and Raes 2004).[4] At a larger scale, an expectation of divergence is present in standard phylogenetic models that treat species as particles and their evolution as a Markov process (Raup et al. 1973). The principle arises too in null models of the evolution of phenotypic diversity (also known as "disparity"), which predict that physical differences among species tend to increase (Foote 1996; Gavrilets 1999; Ciampaglio, Kemp, and McShea 2001; Pie and Weitz 2005; Erwin 2007). For complexity, the principle underlies the notion of duplication and differentiation of parts (Gregory 1934, 1935), including genes, considered as parts. And it is traceable historically to Herbert Spencer's (1900) notion of the "instability of the homogeneous." Spencer argued that the parts of individuals in a lineage should tend to become more different from each other as they accumulate heritable accidents.

What is new, what the ZFEL offers, is a recognition of the unity among these cases, of the common thread that runs through standard thinking about them. The ZFEL makes the common principle explicit and gives it a name. The ZFEL also turns out to have real consequences for research. It is testable, and as we will argue later, it opens new research avenues. In particular, it reconfigures the long-standing problem of the origin of and rise in organismal complexity. As we will see in chapters 5 and 7, it suggests that the real puzzle in evolution is not why organisms are so complex but why they are not more so.

More generally, what the ZFEL offers is a gestalt shift for evolutionary biology, a radical change in our view of what is pattern, and therefore needs special explanation, and what is background. In the standard view of evolution, increases in most variables are understood to require a force, such as natural selection. In the ZFEL view, increase is the background condition, with natural selection in the role of superimposed force, augmenting or opposing the background increase.

A Newtonian Analogy

We propose that the role the ZFEL plays in evolutionary theory is analogous to inertia in Newton's first law. Inertia—lack of change—is the default, or "natural," state of velocity, the background against which gravity and other special forces act. The first law says that, if no force acts on an object, its velocity will remain constant. It is deviations from constant velocity that require forces. Analogously for diversity and complexity in biology, it is deviations from the increase predicted by the ZFEL that require forces. Paradoxically, our familiarity with Newton's laws makes the claim here somewhat counterintuitive. As Newtonians, we are comfortable with the idea of constancy as the null expectation. (That is, constancy of velocity, of course, not constancy of position, unless velocity is zero.) Indeed, so deeply ingrained is the Newtonian paradigm that we sometimes accept the following as a gloss of Newton's first law: if no force, then no change. In contrast, the ZFEL says that, in evolution, the expectation in the absence of any forces is change. If no force, then change.

A Law of Evolution

We have chosen to call the zero-force principle a law for two reasons. First, it is true everywhere and always, in all evolutionary systems with variation and heredity. That is, it applies equally to all evolutionary systems on Earth, past and present, and to all evolutionary systems that may exist, or may have existed, elsewhere. If, as Darwin said, "natural selection is daily and hourly scrutinising, throughout the world, every variation" (Darwin [1859] 1964, 84), then the ZFEL is daily and hourly tending to increase variation, throughout this and all other life-bearing worlds. It is universal, we claim. Second, the ZFEL is not analytic. It is not true as a matter of logic or mathematics, as is biology's so-called Hardy-Weinberg law. Rather, it is synthetic, making an empirical claim about the way the world is. The world could in principle have been otherwise. However, the law is different from most other synthetic generalizations in biology that have been called laws, such as Mendel's law of independent assortment (alleles for different characters segregate independently), and different too from any of the many "rules" that are sometimes called laws, such as Cope's rule (that body size increases on average). The ZFEL is not an empirical generalization that arises from other contingent facts of biology or from observation of the pervasiveness of some phenomenon. We have not examined the data on diversity and complexity over the history of life and discovered there an increasing tendency.[5] Rather, the ZFEL arises from the contingent properties of variation in nature, properties that are the formal domain of probability theory. (See chapter 6.)

Law and a Gestalt Shift

The two main points of the last few sections are worth a summary restatement. We are proposing both a new law and a gestalt shift. The law is a universal tendency for diversity and complexity to increase. And the gestalt shift places this tendency in the background, moving the effect of natural selection and various constraints on diversity and complexity to the foreground. Our experience is that those who pay attention to the law but overlook the gestalt shift are sometimes confused by our claims. In particular, they may imagine we are claiming that diversity and complexity *must* increase and that the ZFEL is the cause whenever it does, claims that are instantly refuted by the many instances of decrease known from the history of life and the many cases of diversity and complexity increase known to have other causes, such as selection. In fact, however, what we claim to have identified is a background tendency, one that acts everywhere and always but that may be overcome at any turn by foreground forces (such as selection) and constraints. To help the reader keep the structure of the argument in mind, we recall it, using various analogies, throughout the book—we hope not so often as to test the reader's patience.

Diversity and Complexity

Our usage of "diversity" is conventional, but that of "complexity" is not. Here "complexity" just means number of part types or degree of differentiation among parts (McShea 1996). This choice will sound odd because in colloquial usage complexity means so much more, connoting not just part types but functionality, sophistication, and integration, among other things. We call complexity in the parts-and-differentiation sense "pure complexity," to distinguish it from the much broader "colloquial complexity." Some will find the notion of pure complexity disconcerting on account of the severing of any connection with function and natural selection. The central questions in biology have centered on function, on what the parts of an organism are *for,* and how they interact to enable the organism to *do* things, behaviorally or physiologically. But in our usage, even functionless, useless, part types contribute to complexity. Even maladaptive differentiation is pure complexity. We understand the objection. And we have two answers to it. First, we agree that function is important. But this book is about something else. The questions we raise have to do with "How many?" rather than "What for?" They have to do with "How differentiated?" in a sense that is independent of "How capable?" In effect, what pure complexity does is enable us to ask the same questions about organismal structure that diversity lets us ask about ecological structure. Diversity too is a "how many" concept. How many types of individuals? How many different species? To ask these "how many" questions about diversity and complexity, definitional independence from function is essential.

Second, a notion of pure complexity, independent of function, is essential precisely so that one can ultimately address the relationship between complexity and function, between complexity and natural selection. One could not, for example, ask whether complexity is favored by natural selection, whether complex structures are more functional than simpler ones, on average, if one's notion of complexity had the effects of natural selection—that is, function—built into it. In investigating a relationship between *A* and *B,* it is helpful, to say the least, to define *A* and *B* so that they are conceptually independent of each other. We discuss this further in chapter 4.

A consequence of our use of a function-free notion of complexity is that there will initially be little connection between what we are talking about and "complexity" as it has been used in most of the evolutionary literature. Peppered throughout that literature are references to "organized complexity" and "adaptive complexity," as well as to the "information content" and the "computational power" of organisms, with an assumed connection between information or computation and complexity in some sense. Often these terms arise in discussions of what is taken to be an obvious directionality in evolution, sometimes called "evolutionary progress," the rise in "complexity" from bacterium to human. The concern in this literature is clearly with complexity in the colloquial sense, and we will return to colloquial complexity or something like it (in chapter 7) in order to say something about how it might arise in evolution. But, as we hope will emerge, understanding its poorer cousin, pure complexity, is a necessary first step. In the meantime, we ask readers to hold in suspense everything they think they know about complexity in biology, including its possible increase in evolution. We will be offering a new view in which the colloquial notion and the possible trend emerge in a new light.

It is also worth mentioning here at the outset that our treatment shares the word "complexity" with a field called "complex systems," a hot area in the past twenty years. Complex systems are usually taken to include not only organisms but also markets, technologies, social and political organizations, many-body physical systems, and computer programs with problem-solving abilities like cellular automata, NK networks, and neural nets. We want to be clear here that our understanding of complexity is different, and that there is no connection between our project and these various complex-systems research programs, at least no direct connection.[6] Complexity in our sense, pure complexity, is not a function of the nonlinearity of interactions in a system, the sophistication of a system, or a system's ability to survive, reproduce, adapt, compute, or think. Pure complexity is not connectedness or integration. It is not the length of the shortest description of a system or of the algorithm for generating it. It has nothing to do with the amount of energy a system uses or how it uses it. In this book, the phrase "pure complexity," or just "complexity" alone and unmodified, always means number of part types or differentiation among parts. And nothing more.

Hierarchy

In the chapters that follow, hierarchy will emerge as central in our understanding of the ZFEL. We understand hierarchy in the sense of physical nestedness.[7] Higher levels are composed of lower-level parts. Aggregates of lower-level units physically constitute higher-level entities. As a preview of the role of hierarchy, consider these two points. First, the ZFEL applies at all levels where there is heritable variation. It predicts increase in diversity and complexity of genes, macromolecules, organelles, cells, tissues, organs, individuals, groups, populations, species, clades, or higher-level units.[8] Second, the ZFEL applies to each independently. Indeed, in principle, the ZFEL could manifest itself differently at each level. This is an important point. As we will see, measures of complexity and diversity are level relative. For instance, organismic complexity might be measured in terms of differentiation among cells or, in the discrete case, number of cell types. But cellular complexity might be measured as the number of types of cell structures, organelles, and such. Thus, it is perfectly coherent to say, for example, that in evolution, multicellular organisms became more complex (more cell types) while their component cells became simpler (losing certain cell structures, perhaps favored by selection for specialization) (McShea 2002). Similar remarks could be made about diversity. A clade might become more diverse (more species) even while its component species decreased in diversity (less variation within each species). Given all this, our hierarchical approach is not just convenient, it is necessary. No single-level account, no reductionistic account, could capture the phenomena. Thus, in our hierarchical view, the genetic or macromolecular level—deemed conventionally to be central to understanding evolution—is just one among many and is not privileged in any way. As a consequence of our apostasy on this matter, genes and macromolecules, although discussed in what follows, are not mentioned explicitly in the ZFEL and are not central.

Diversity Is Complexity, Complexity Is Diversity

Diversity and complexity might seem like an odd pair. In conventional usage, one is a much-studied and well-understood property of populations or taxa, and the other is a poorly studied and barely understood property of organisms. So it might seem strange that the same principle would apply to both. The explanation is simply that both diversity and complexity are aspects of variance. As mentioned earlier, diversity can be understood either in its continuous sense, as degree of differentiation among organisms or taxa, or in its discrete sense, as number of types of organisms or taxa (e.g., number of species). In either case, it is a function of the amount of variation among organisms. Pure complexity—whether understood as degree of differentiation among parts or as number of part types—is likewise a variance concept. It is variance among parts within an organism rather than among organisms. Thus, the ZFEL says that variation occurs, and in systems with inheritance it tends to accumulate, with the result that variance increases. And it will do so whether the system consists of a set of organisms or a set of parts within an organism.

But the relationship between diversity and pure complexity is closer than that. They are really one and the same thing, considered from hierarchically adjacent vantage points. That is, the diversity of a system at level N is just its complexity at level $N + 1$. For instance, diversity at the cellular level = complexity at the organismic level. An organism with a great diversity of cell types is a complex organism. Moving up a level, diversity at the organismic level = complexity at the group level. A group of organisms that is diverse can be said to be a complex group. This last is not ordinary usage of the term "complex," of course, and it will sound odd to most biologists.[9] But that is because most think of complexity as a compound notion implying both number of part types and functional organization. However, recall that in our technical treatment here, we are considering only the function-free aspect of complexity, pure complexity (see chapter 4). Thus, the identity between complexity and diversity follows.

In the following chapters, we consider diversity and complexity separately because they have been treated separately in the literature, and different issues have grown up around them. So the ZFEL needs to be framed differently for each of them. However, as we hope will be clear, the argument is essentially the same for both.

Diversity and Complexity, Not Adaptation

It should be obvious already that this treatment differs from the vast majority of empirical and philosophical works in biology in that the mission is *not* to understand the origin and evolution of adaptation. Our concern is with diversity and complexity, which natural selection can act upon but which, we claim, are also subject to a tendency independent of selection. We have done, and will do, everything in our power to convince readers that we are serious about this unusual focus. But given the adaptationist bent of biology since Darwin and the power of traditional formulations to straitjacket thinking, we expect that some will occasionally find themselves confused about the nature of our project. To these readers, we first apologize for the limitations of our rhetorical skills. And second we encourage them, when faced with apparent incongruity, to rethink what we are saying with our special focus in mind: the ZFEL is about a particular property of biological entities, the amount of variation in them, and how the amount of variation is expected to change in evolution. As will be seen, the ZFEL has consequences for adaptation. But it is not a claim about adaptation. Returning to our analogy with inertia and Newton's first law, our claim is that just as the Newtonian needs the First Law to understand the effects of special Newtonian forces, so the evolutionary biologist needs an understanding of the ZFEL before investigating adaptive evolution.

The ZFEL and the Second Law of Thermodynamics

Based on what we have said so far, some will be poised and ready to make a leap, from the notion of accumulation of accidents to the second law of thermodynamics (Pringle 1951; Brooks and Wiley 1988; Collier 1986,

2003). We advise readers against this, for their own safety. We are concerned that on the other side of that leap there may be no firm footing. Indeed, there may be an abyss. First, we think the foundation of the ZFEL lies in probability theory, not in the second law or any other law of physics. And second, our notions of diversity and complexity differ fundamentally from entropy, in that entropy, unlike diversity and complexity, is not a level-relative concept. (We explain these claims further in chapter 6.) Still, some work in recent decades on the application of the second law to biology has been inspirational (especially Wicken 1987; Brooks and Wiley 1988; Salthe 1993), and we gratefully acknowledge the intellectual debt.

Outline of the Book

Chapter 2 explains our understanding of randomness, as it relates to the ZFEL, and addresses the in-principle problem of how random processes can create a directional tendency. It also explains how we are thinking about hierarchy and constraint. Chapter 3 explains how we are using the term "diversity" and how the ZFEL applies to diversity. Chapter 4 does this for "complexity," giving an extended discussion of our choice of this word and explaining our understanding of the notion of a "part." In each of these two chapters, we offer a key piece of the evidence that supports the ZFEL, and then chapter 5 reviews the evidence more broadly. Chapter 6 treats certain philosophical issues that arise, including a possible theoretical foundation for the ZFEL. And chapter 7 discusses some of the implications of the ZFEL for biology.

2

Randomness, Hierarchy, and Constraint

We begin with three preliminary points about the role of randomness in the ZFEL. First, we raise and then solve an apparent puzzle arising in connection with the ZFEL, one that will have occurred to some readers. It is the problem of how a directional process—the increase in diversity and complexity predicted by the ZFEL—can arise from random variation. Second, we address the question of whether the ZFEL is true for all starting conditions. Obviously, when all pickets in a fence are the same, any variation that arises can only increase the variance among them. The same is true when all individuals in a population are the same or when all parts within an individual are identical. Given a zero-variance starting condition, there is nowhere for variance to go but up. But what will be the effect of random processes when the starting condition is inhomogeneous? Is the ZFEL true even for populations that are already diverse or individuals that are already complex? Here we argue that it is—in other words, that in the absence of forces and constraints, diversity and complexity are expected to increase indefinitely. Third, we explain how we are using the term "random" and make the point that, given this understanding, randomness is relative to a hierarchical level of interest. Thus, events that are determined at one level may nevertheless be random "with respect to" each other and thus are effectively random at the next level up. With randomness understood in this with-respect-to sense, the ZFEL can be true at one level, even though deterministic processes govern at lower levels. This point is essential to the discussion in later chapters.

We close this chapter with a discussion of how we intend the phrase "absence of constraint" to be understood in the special formulation of the ZFEL. Clearly, constraints are never entirely absent from any system. As we explain, some constraints are understood to be constitutive of a system, while others are imposed. It is the imposed constraints that are taken to be absent in the special formulation.

Random Processes and Directional Predictions

Unlike inertia, the ZFEL is a probabilistic process. Now it is easy to see that a probabilistic process can make directional predictions. Selection is probabilistic, and it can make directional predictions, for example, selection for increased body size. A random walk driven by flips of a biased coin is probabilistic and directional. In both cases, the directionality is possible because of a bias. But the ZFEL arises from an *unbiased* random process, and yet it predicts directional change: increasing diversity and complexity. How is this possible? The answer is that diversity and complexity are level-relative properties that attach to a level higher than that of the random underlying process. That is, random diffusion at one level can be manifest as directional increase (in complexity or diversity) at the next higher level. Thus, the ZFEL requires a hierarchical perspective.

To see this, consider a particle moving in a two-dimensional space. Let the y-axis represent time measured in discrete steps and the x-axis represent space (see the axes in fig. 2.1). The particle starts at the origin and at t_i moves one step either to the right or to the left. The probability that it moves to the right at that time is 0.5, as is its probability of moving left. For each n, the particle's movement at t_n is governed by the same rule. Where on the x-axis will the particle be at, say, t_4? The expected outcome is that it will be at 0, but that is but one of five possible outcomes—the others being $x = 4$, $x = 2$, $x = -2$, and $x = -4$. Given the probabilities associated with each possible transition at each time, it is easy to calculate the probabilities associated with each of the five possibilities. The expected outcome, $x = 0$, has a probability of 0.375. Here are the probabilities of the other possible outcomes: $\Pr(x = -4) = 0.0625$; $\Pr(x = -2) = 0.25$; $\Pr(x = 2) = 0.25$; and $\Pr(x = 4) = 0.0625$. Thus, the

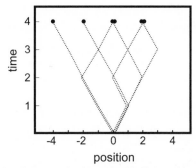

FIGURE 2.1 The increase in variance over time in an ensemble of six particles.

likelihood that the particle will be in the positive region of x-space is identical to the likelihood that it will be in the negative region. There is no directional tendency. And that is true no matter how far we extend this process in time.

Now consider an ensemble of such particles, as in figure 2.1. Ensembles, like particles, have properties, and one property of an ensemble is the mean. Notice that the mean is a property of the ensemble but not of individual particles. No individual particle has a mean at a given time. And in that sense, the mean is a property at a higher level. Also like the particles, the mean has an expected value at a given time, an expected location or value in the one-dimensional space of the model, the horizontal axis. In the model, the expected value starts and remains at zero. That is, the expectation is no directional change in the mean. Consistent with intuition, random change of lower-level particles in the space of the model does not produce any directional change in a higher-level property, the mean, represented in the same space.

Another ensemble property is the dispersion, or the variance in position of the particles. The model does make a directional prediction about variance, namely that it is expected to increase in each time step. Like the mean, the variance at a given time is a higher-level property, a property of the ensemble but not of individual particles. No individual particle has a variance at a given time. However, unlike the mean, a variance at a given time cannot be represented as a point in the space of the model. In other words, it cannot be represented by the descriptive tools adequate at the particle level. One way to say this is that the variance is a property of a higher *order*. And for higher-order properties, intuition is silent. We have no reason to think that random movement in the space of the particles should translate in any direct way to behavior of higher-order properties in a very different space. In particular, there is no reason why purely random behavior of particles should not produce an increase in variance.

Variance Increases Indefinitely

Suppose we run the model illustrated in figure 2.1 over a longer time span, measuring the variance at each time step using some appropriate metric, say the standard deviation. The top graph in figure 2.2 shows the result. The standard deviation trends decidedly upward. Interestingly, despite this upward trend, the graph shows occasional downward wobbles. This is not unexpected, because the ZFEL prediction of increase is probabilistic. By chance, dispersion *can* decrease, in that points could move closer together. And in runs of the model starting with small sets of points, this happens moderately frequently, and not infrequently even with larger sets. Notice, however, that this would not be true of a more realistic model with higher dimensionality. Even if the points by chance become less dispersed in one or a few dimensions, they are likely to become more dispersed in most. In a multidimensional space, the increase in overall variance will be quite reliable.

The bottom graph in figure 2.2 shows the average trajectory of the standard deviation over 1000 runs of a one-dimensional model (with error bars showing one standard deviation on either side of that trajectory, in effect one standard deviation of the standard deviation). And it reveals something that some may find surprising. The initial rise in diversity or complexity is not surprising. Because all points start at the same place, dispersion can only increase. What may be surprising is that dispersion continues to rise even later, when the points have become quite dispersed. Even though the starting condition is high diversity or high complexity, the expectation is further increase. Analytically it can be shown that this is a square-root curve and that it does not asymptote. Variance rises indefinitely.

The ZFEL claim is that variance increases indefinitely in the absence of limits or external forces, not just for a standard deviation metric, but for any intuitively reasonable measure. We could use the statistical variance, for example, or we could use something like the average absolute difference between pairs of points. Other measures will produce trajectories of different shapes, of course. The statistical variance will produce a straight line rather than a square-root curve, for example. But almost any measure of average dispersion will increase monotonically over a number of independent runs and will do so indefinitely.

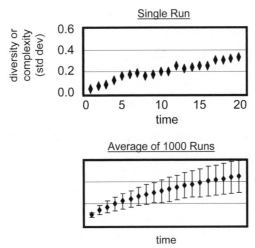

FIGURE 2.2 *Top:* Trajectory of diversity or complexity in an ensemble of points in a single run of the model. *Bottom:* The average trajectory over 1000 runs. Error bars show one standard deviation of our variance measure, which is also a standard deviation.

Measures can be devised that appear to contradict the ZFEL. For example, one could carve the horizontal axis into discrete bins and measure the variance as the number of different bins occupied. And in that case the variance is expected to rise until every point occupies its own bin and then to increase no further. But what has happened here is that the decision to recognize only the number of bins occupied, irrespective of how far apart the points are, places a mathematical upper limit on the amount of variation that the metric is able to record. One can think of the limitation as a constraint imposed by the metric or by the number of points. And the appearance that the variance has asymptoted is the result of that limit. We discuss the implications of the use of discrete metrics further in chapters 3 and 4.

In sum, what is modeled here is a diffusion process. The point is that, unless some constraint or force is imposed, diffusion should drive variance upward indefinitely.

Randomness in the "With-Respect-To" Sense

Our argument in the following chapters hinges critically on a hierarchical understanding of randomness. In the example above, each particle can be understood to be governed by what might be called true chance, perhaps at the quantum level, so that its movement left or right is understood not to be the product of any deterministic forces at all. But this view is optional. Each particle can equally well be understood as governed at each point in time by a set of deterministic forces, perhaps complexly configured in space and time so as to produce leftward or rightward movement half the time. To the extent that the set of deterministic forces acting on each particle at each time is independent of the set of forces acting on other particles, the behavior of each particle can be said to be random *with respect to* all other particles. In a group of people flipping coins, each person produces a series of heads and tails that is random with respect to the series produced by every other person. And this is true even if we understand each coin toss to be completely deterministic.

From the perspective of the ZFEL, the two kinds of randomness—true randomness and randomness in the with-respect-to sense—are equivalent. Whether the movement of each particle is truly indeterministic or whether it is deterministic but independent of all other particles, the expected result is ever-increasing dispersion, that is, increasing variance. Consider this. Take a snapshot of all of the people on a crowded city street corner at some moment in the middle of the day. Then find these same people 10 minutes later. Find them again 20 minutes later, and then 30 minutes later. With the passage of time, they will become increasingly dispersed, or in other words, the variance in their locations will increase. And this is true even if the trajectory of each person is the deterministic outcome of his or her plans for that afternoon. One is on her way to her office. Another is walking his dog. A third is going grocery shopping. And so on. To the extent those motivations and plans are different from and independent of each other, the individual movements are random *with respect to* each other. And dispersion at the higher level—the greater variance in location of the group—is the expected outcome of randomness in the with-respect-to sense at the lower level. This is the principle underlying the ZFEL.

Or consider a biological example. Suppose that several populations in a species are independently undergoing drift. The ZFEL says that they will tend to diverge. That is, diversity within the species will tend to increase (where diversity is here understood as between-population diversity). But notice that the same thing would happen if each population were evolving under tight selective control but independently, with a different selection pressure acting on each population. In a plant species, selection in one population might favor increased resistance to some local environmental toxin, while in another population selection might favor drought resistance. Each population changes deterministically, but they change randomly with respect to each other (Eble 1999). The

ZFEL prediction is the same whether the underlying cause is drift or directed changes that are random in the with-respect-to sense. This second sense of randomness works at higher levels too. One species in a genus of sea urchin might be under selection for greater spine length while another is under selection for greater mobility. The selective forces are random with respect to each other, so the ZFEL expectation is that the species will become more different from each other. The genus will become more diverse. We will say more about this sort of ZFEL-driven divergence in chapter 3.

Imperfect Independence

Seemingly overlooked in our model above is the possibility that movements among the points could be correlated. In the human-dispersal analogy above, if a transportation strike has just been announced as imminent, the people may all move in concert in the same direction, perhaps to the nearest bus stop, under the influence of a common desire to get home before the strike begins. In biology, developmental mechanisms commonly produce correlated changes among parts, for example, the correlated changes in parts in serial structures or the correlated change in sizes of parts that accompany evolutionary changes in body size.[1] As a result of these correlations, complexity in the sense of differentiation among parts need not increase. Among species, there is selection-driven convergence on a common phenotype, as occurs in mimicry or the many cases of evolutionary convergence. Diversity in the sense of differentiation among species need not increase. It might seem that at least some subset of these might produce results that falsify the ZFEL.

However, first, recall that the ZFEL says that diversity and complexity will increase on average in the absence of forces and constraints. And correlations among parts are usually understood to be the result of developmental constraints. Mimicry is the result of selection for similarity (i.e., selection against diversity of form). Convergence is the result of selection in a common environment for a particular adaptive phenotype. And where constraints and selection act to maintain or produce similarity, the special formulation of the ZFEL does not apply. The general formulation says only that there is a *tendency* to differentiate, one that in these cases has been overcome.[2]

Second, although not essential to the truth of the ZFEL, it is worth pointing out that all correlations are imperfect. No two objects in the universe are perfectly correlated with each other. Each experiences at least slightly different forces than the other, minimally on account of occupying a slightly different location in space. Within an organism, two similar parts may find themselves in, say, a common developmental growth field. But they will nevertheless respond at least somewhat differently to it, owing to their somewhat-differing locations within the field. Further, even ignoring that source of difference, each is expected to respond uniquely to the same force on account of the inevitable minor differences between it and the other (Spencer 1900). Each part in an organism has a composition that is at least a little different from its neighbors, and therefore its response to the same external influence will be to some degree unique. Likewise, the combined biotic and abiotic environment of each individual and species is to some extent unique, however minimally. And to the extent that parts, individuals, and species are unique and are affected by unique factors, they are expected to change randomly with respect to each other and, therefore, the ZFEL says, to become more differentiated.

Constraint

The special formulation of the ZFEL speaks of "absence of constraint." To explain how we intend this to be understood, we need to make two distinctions, the first between *real* and *effective* constraints. Consider a genome of length L, where L is the number of nucleotides. The total number of possible genomes is 4^L, because there are four possible nucleotides at each position. For most genomes, the 4^L limit is a real constraint, in that it represents a real upper bound on diversity that cannot be crossed, given the initial conditions of four nucleotides and fixed L. But it is usually not an effective constraint. In the human genome, for example, the 4^L ceiling is much higher than the population size of our species, now and in any imaginable future, so that even though every individual has a unique genotype, diversity is *effectively* unconstrained. The limit has not been and will never be actually encountered. In the special formulation of the ZFEL, the phrase "absence of constraint" refers only to effective constraints. Constraints that are real but not effective do not limit diversity or complexity and do not need to be absent in order for the ZFEL to operate.

Still, the notion of a system in the "absence of constraint" sounds a bit strange. A system that is literally and totally without constraint would be one without properties, which is unimaginable. However, we do not intend this phrase to be understood as all-encompassing in the special formulation of the ZFEL. For one thing, the only constraints that are taken to be absent are those on diversity and complexity. But in addition, there are some constraints that we take to be *constitutive* of the system and others that we take to be *imposed*. This is the second distinction we need. The special formulation of the ZFEL excludes only imposed constraints.

A constitutive constraint is one that is relegated to the explanatory background, either of necessity or for heuristic purposes. For the special formulation of the ZFEL, there are two necessary constitutive constraints: reproduction and heredity. We call these constitutive because they are fundamental to life, wherever and whenever it occurs. They are constraints that cannot, even in principle, be removed or avoided. In contrast, imposed constraints on diversity and complexity are any *beyond* the constitutive ones, such as those arising from logic, mathematics, physics, evolutionary history, development, material properties, and so on. Imposed constraints are always removable, or at least avoidable, in principle. Consider a developmental constraint. In marine molluscs, there is a constraint on diversity imposed by planktotrophic feeding, a life habit in which larvae disperse widely, discouraging the isolation of populations that disposes a taxon to diversification. But this constraint is avoidable by switching to a nonplanktotrophic life habit, in which larvae settle and develop more locally. What about mathematical constraints? The complexity of limb pairs in tetrapods is limited to two different types, a consequence of the mathematically inescapable truth that in a system of N elements, in this case two, the maximum number of different types is equal to N. But this constraint could be avoided, for example, if a tetrapod were born with three limb pairs. Imposed constraints are those that are breakable or avoidable. And it is these that must be absent in order for the increase in diversity or complexity predicted by the special formulation of the ZFEL to be observed.

In particular contexts, there will be some constraints that are removable in principle but can still be treated as constitutive for heuristic purposes. In the particle example, the setup implicitly included a constitutive constraint: that each particle continue to exist from one time step to the next. This was a necessary constitutive constraint. It would be hard to think about a system of diffusing particles if they were winking in and out of existence. But the particle example also included the heuristic constitutive constraint that each particle move exactly one unit in each time step. This constraint could easily have been relaxed, allowing the particle to move, say, 1000 units in each time step (with the result that the variance would increase much more rapidly).

In biological systems, it is often helpful to think of certain constraints as constitutive if we have reason to think they are constant or reliably present on some long timescale. For example, in the genome example above, both L and 4 are constraints on diversity (albeit not effective ones, but never mind that for now). However, the timescale on which the number of possible nucleotides might increase is considerably longer than the variational timescale for L. Nucleotide options have remained essentially unchanged, it is generally thought, since early in the history of life and probably will not change for a very long time in the future, if ever. They are what is commonly called a frozen accident. And so it is convenient to think of 4 as a constraint that is constitutive, not of life in general as are reproduction and heredity, but of life on Earth over most of its history. In contrast, genome length, L, varies a great deal on this timescale and is therefore treated as an imposed constraint. So for life on Earth, when the special formulation of the ZFEL says that diversity will increase in the absence of constraint, the implication is that this will be true whenever L (and any other imposed constraint) is not limiting, that is, not effective.

Importantly, when the distinction is made for heuristic purposes, the ZFEL does not hinge on how it is made. That is, the special formulation of the ZFEL is not falsified if, in the absence of all imposed constraints, diversity or complexity fails to increase. It may be simply that we need to treat as imposed some constraint previously thought of as constitutive. Suppose that in the absence of all identifiable imposed constraints, some small-genome organism with a huge population size actually encountered the 4^L constraint on diversity. The ZFEL would not be falsified. It might simply mean that we need to treat the 4-nucleotide constraint (as well as the L constraint) as imposed. And clearly if *that* constraint were absent—if we allowed a fifth nucleotide—genomic diversity would again increase.

A final point. The conditions we have set out for the special formulation of the ZFEL may sound rather onerous, that is, not likely to be met in any real biological system. Constraints beyond reproduction and heredity are always present in real biological systems. In particular, every property or dimension of the system that is fixed, or bounded in its range of variation, is by definition constrained, many of them by effective constraints. Do we demand that all constraints that limit variation, beyond reproduction and heredity, be removed in order for the ZFEL to be expressed? The answer is no. The ZFEL will be expressed in any and every dimension of the system in which imposed constraints are absent (or, if present, then not effective). All that is required, minimally, is one such dimension.

We are now in a position to explicate more precisely the "absence of constraint" clause of the ZFEL. The ZFEL says that, given a system with the constitutive constraints of reproduction and heredity, whenever effective, imposed constraints are absent in any dimension of interest (along with, if necessary, any constraints that are being treated as constitutive for heuristic purposes), diversity and complexity will increase, on average, in that dimension.

3 Diversity

Consider a population of initially identical organisms. Suppose they are also reproducing asexually (although this is not critical). In the absence of constraints, natural selection, or any other force, mutation will cause individuals in the next generation to differ from each other. And in each subsequent generation, the expectation is that differences among individuals will increase. More precisely, at any given time, for any character with some measurable dimension, a population of individuals will have some distribution in that dimension. The ZFEL says that at some later time, in the absence of constraints or intervening forces, the variance of that distribution will tend to be higher. For continuous characters, the distribution is expected to spread, with the upper and lower tails diffusing up and down respectively. For discrete characters, the increase may take the form of a diffusion and redistribution of characters among existing states, or it may occur via the addition of novel states, extending the number and range of states realized, or both. All are increases in diversity.

That is what the special formulation of the ZFEL says. In the absence of constraints and forces, random variation arising in parents makes offspring more different from each other than were the parents, on average. If we instead look to the general formulation, it says that there is a *tendency* for diversity to increase whether or not constraints or selection are present. Constraints and selection may block increase, preventing the ZFEL from manifesting itself. But the tendency to increase is present nonetheless.

The ZFEL should operate over a wide range of scales. It predicts increasing diversity of a population (where the components are individuals), as just discussed. But it also predicts increasing diversity of a species (made up of populations), of a clade (composed of species or lineages), and of all life (composed of clades). Further, it predicts that populations, species, and clades should become more diverse in every property—morphological, physiological, and behavioral—and at every level of organization, from macromolecules and cells to tissues and organs (if multicellular).

Notice the qualification in the ZFEL. The special formulation says that, in the absence of selection and constraints acting on diversity, it will increase, *on average*. This means that it *can* decrease, by chance, given the right chance combination of mutation and deaths and, if sexual, the right combination of mate choices and recombination events. A diverse population of individuals ranging from short to tall could by chance give rise to a less diverse generation of all tall individuals. But probably it will not.

About Diversity

As the word is used colloquially, and often technically, "diversity" refers to number of discrete species or, more generally, number of discrete taxa. A diverse rainforest is one with many species. But the essence of diversity is variation, discrete or not. If all organisms on Earth were distributed continuously in every dimension, with all intermediates present and no discernible taxa, we would still recognize life's enormous variety, its diversity. Indeed, when discrete taxa cannot be identified, as with ring species or phyletic transformations in the fossil record, we still recognize diversity (even assigning species names to phenotypes that are sufficiently different, despite the arbitrariness of doing so in a continuum). Further, even where species are discrete, diversity may still be a continuous variable, as recognized in the notions of "morphological diversity" or "phenotypic diversity" or by a term of art in paleobiology, "disparity" (Foote 1997; Erwin 2007). In paleobiology, the disparity of a genus might be the statistical variance of some univariate (say, body size) or multivariate phenotypic measure among species in the genus.

In sum, a population with many different kinds of individuals, or a clade (or a biota) constituted by many different taxa, is diverse. And a disparate population with individuals that are highly differentiated phenotypically, or a clade with highly differentiated species (or higher taxa), is also diverse.[1]

Abstract definitions are all very well for certain purposes, but some will reasonably wonder how we are operationalizing diversity. The answer is that we can and do operationalize ad hoc, choosing diversity measures as needed to suit the context or data at hand. The reason we can do this is that the ZFEL predicts an increase in *all* intuitively reasonable measures (at least all continuous measures; see below). For example, the measures that have been used in the paleobiological literature on disparity include not just the statistical variance of some phenotypic measure but the range of variation, the average pairwise difference among individuals or taxa, the average distance from the phenotypic centroid, and others (reviewed in Erwin 2007). The ZFEL predicts that, in the absence of selection and constraints, all of these will increase.

In this book, when we use the term "diversity" generically, without reference to any particular biological data set, we are referring to both discrete and continuous variation, number of taxa and disparity. But we need to point out that the continuous sense, disparity, is more general.[2] And further, technically, it is only to disparity that the special formulation of the ZFEL applies without qualification. To see why, imagine a set of asexual organisms distributed along some axis of variation. And suppose that we adopt a discrete measure of diversity, dividing the axis into bins, corresponding to species. The ZFEL predicts that, in the absence of selection and constraints, the number of bins occupied (the number of species) will increase over time. However, if the population is not growing, diversity—the number of bins occupied—will rise only until every individual occupies a unique bin, until every individual is a unique species. And it will increase no further, apparently contradicting the ZFEL. But in fact there is no contradiction. The apparent leveling of diversity is an artifact of the choice of a discrete measure, a measure that has a built-in constraint and that as a result is insensitive to further increases in disparity. Information-theoretic measures of diversity, such as the Shannon index, have this same insensitivity. Alternatively, one could say that the constraint is population size, which limits the degree to which a discrete measure can capture true diversity. In contrast, continuous measures of diversity, such as the statistical variance, range of variation, and so on, are not insensitive in this way. And that is why we call diversity in the continuous sense—disparity—more general.[3] In real populations, of course, this limitation is a problem only in principle, because the limiting case in which each individual is a unique species is never even approached.[4] But we need to say that, technically, the no-constraint condition for the special formulation of the ZFEL is met only when the constraints imposed by a discrete metric or population size are distant, that is, when they are not effective. Of course, the conditions for the general formulation are always met, for both senses of diversity. A *tendency* for diversity to increase will be present even if every individual is a unique species.

Diversity as a Level-Relative Concept. It is well appreciated, but worth stating anyway, that diversity can vary independently among hierarchical levels, that diversity at one level is at least conceptually independent of diversity at another. The diversity of a population or subspecies, understood as, say, the number of different sorts of individuals within it, is independent of the diversity of the species that contains it, understood as the number of different subspecies. For example, the first value could be high while the second is low, or vice versa. The same is true across the hierarchy spectrum. The diversity of a family, measured, say, as the disparity among the genera that constitute it, is independent of the disparity of a genus within that family, measured as the differentiation among its component species. The reason is simply that diversity is a level-relative concept. From the ZFEL perspective, this independence is important. The ZFEL predicts not only that diversity at all levels will increase, as noted earlier, but that increase at one level is not tied to increase at another. The ZFEL tendency for a population to become more diverse could be blocked, perhaps by selection, and yet the ZFEL still predicts that the species of which it is part will become more diverse. Extinction of genera might be reducing disparity at the family level even while the ZFEL drives up disparity among the species within the surviving genera. As a consequence of this independence, there is no privileged level of analysis, no single level at which ZFEL-driven diversity increase is expected to be manifest, and therefore no single level to look to when it comes to testing.

Heredity and Reproduction

The ZFEL requires that some of the variation that arises be heritable, but it is indifferent to the mechanism of inheritance. It can be DNA mediated or some epigenetic process. And there is no requirement that *all* variation be heritable. Where it is not, the ZFEL simply does not apply.

The requirement for heredity is unproblematic for the ZFEL in that apparently all reproducing organisms exhibit it to some degree. Of course, there are processes tending to limit heredity or undermine it, such as DNA error correction, which undermines the inheritance of mutations. In the ZFEL view, such processes are treated as either forces or constraints. DNA error correction can be thought of either as a constraint or as the result of a force, selection.

In biology, heredity by definition requires reproduction, and therefore implicit in the heredity requirement of the ZFEL is a requirement that organisms reproduce. This too is unproblematic, in that reproduction is likely a universal property of life.[5]

Constraints and Contrary Tendencies

In this section, we consider the various constraints and contrary tendencies—including selection—that act against the ZFEL tendency in evolution, that is, the factors that reduce or limit diversity.

Limits on Heredity. Diversity increase has limits arising from mathematical, physical-chemical, and historical-phylogenetic constraints on heredity, in particular on the available storage capacity for differences among individuals. In the genome, this could be the number of nucleotides, number of genes, or total length of all the chromosomes in an individual. In the phenotype, it is the number of parts or characters that can vary among the individuals in a population. More generally, the diversity of a population is limited by the number of structural sites, or the amount of structural material, in which individuals in a population can differ. Another limit is the number of dimensions of variation at each structural site. The number of ways any given part or character can vary is large (e.g., size, shape, composition, metabolic rate) but not infinite. Also, the range of variation possible in each dimension is limited. The availability of only four possible nucleotides is a limit on DNA variation, for example. At a higher level, a single cell cannot have the metabolic output of a nuclear reactor. Finally, there is a diversity limit arising from population size. A population with more individuals can "record" and "store" more differences among individuals than a smaller one. A taxon with more species can record and store more differences among species than one with fewer species.

Absorbing Boundaries. A tendency acting contrary to the ZFEL arises from losses due to absorbing boundaries. The best-known case is genetic drift, which predicts that, in any assortment of neutral alleles at a given locus, all but one will eventually drift to a frequency of zero (given a sufficiently low mutation rate). The resulting loss of alleles is a decrease in genetic diversity. However, we think it misleading to identify drift *simpliciter* as the cause of decrease. In general, drift simply causes frequencies to change, imparting no upward or downward tendency and no tendency for variance to decrease. It is only in the region of state space near absorbing barriers that drift has the effect of decreasing variances. Thus, we think it more accurate to say that drift-plus-absorbing-barriers tends to decrease variances in the regions near these barriers. An analogy may make this point clearer. Imagine a yard in which a number of trees are dispersed. And imagine that the wind blows from each point of the compass with equal probability. It is easy to see that in the autumn, when the leaves fall, the wind has the effect of increasing the variance in leaf location. How could it do otherwise? Well imagine that there is a garage adjacent to the yard, and that during the autumn we leave the garage door open. The back wall of the garage now acts as an absorbing barrier for the leaves. The wind direction in the yard is random, but it can only blow into the garage, never out of it. Thus, leaves collect in the garage, and the variance in leaf position decreases. Clearly, attributing this to the random action of the wind is misleading and incomplete. Likewise, to attribute the loss of genetic variation to drift alone is misleading and incomplete.

Now we have no a priori reason to think that losses due to drift-plus-absorbing-barriers will not overwhelm the ZFEL tendency under a broad range of conditions. Indeed, a major research problem in population genetics is explaining how diversity is maintained in natural populations against "losses due to drift." Importantly, however, such losses do not contradict the special formulation of the ZFEL, because they occur in the presence of a constraint, absorbing barriers. We can clarify further by framing the issue under the general formulation of the ZFEL. The general formulation says that diversity has a natural tendency to increase that constitutes the background against which constraints and forces act. In other words, whenever absorbing barriers act to reduce diversity, they do so by overcoming the ZFEL tendency.

Death and Extinction. From the standpoint of the ZFEL, death and extinction can be understood as either constraints or forces, both having many possible causes, all independent of the ZFEL. And both have the same effect. Death reduces diversity. Extinction reduces diversity. These statements are not quite tautologies. Consider the diversity of a higher taxon, say a genus. If the diversity of the genus is defined as the number of species in it, and if extinction eliminates a species, then diversity decreases

by definition. However, if diversity is understood as disparity and measured as a statistical variance in some morphological measure like body
size, then the loss of a species might or might not decrease it. Loss of a
species with an extreme morphology will have this effect, but loss of one
with an average morphology will not. Still, for diversity understood as
number of different taxa and for many measures of disparity—such as
the range of variation in some character—diversity obviously does drop
as losses mount. The same point applies at a smaller scale, within a
population, to diversity decline resulting from deaths of individuals.[6]

Selection against Diversity. Diversity can be harmful. Given a fit parent,
similarity to that parent must often be the best route to fitness for the
offspring. When this is so, random divergence of offspring from the
parent phenotype, and therefore from each other, will be opposed by
selection. DNA error correction is a mechanism that presumably arose
by natural selection not only to limit variation within an organism during its lifetime but also presumably to limit variation among offspring.
DNA repair mechanisms limit the amount of genetic variation both in
the soma and in the germ line. At higher levels of organization, there
are also evolved mechanisms that buffer or canalize development, stabilizing the phenotype against the vicissitudes of both genetic and environmental variation. And then there is selection against the extremes of
variation, or stabilizing selection, acting both internally (eliminating inviable variants) and externally (eliminating or reducing the reproductive
success of less fit variants). Indeed, there are good theoretical reasons for
thinking that stabilizing selection is ubiquitous in nature. If most mutations are deleterious, and if most effects of developmental noise are likewise deleterious, then most selection should be stabilizing. Certainly,
recent molecular evidence shows that stabilizing (or purifying)[7] selection
has operated strongly on particular genes. The now-famous *Pax6* gene
has changed little over the 500 million years that separates fruit flies
and mammals. Of course, studies of individual genes do not allow us
to quantify the prevalence of purifying selection compared with other
forms of selection and compared with drift. For that we would need
large metastudies of molecular evolution. So far as we know, no such
study yet exists, but that should change in the near future.[8]

Other Constraints and Contrary Tendencies. Linkage keeps genes from
assorting perfectly independently, which constrains disparity. Inbreeding
tends to homogenize genotypes, which also reduces disparity. Horizontal gene transfer has the same effect. Just as sexual reproduction can

increase diversity within a population and species, asexual reproduction can have the opposite effect.

Net Effects. Most of these factors probably do not limit the ZFEL much. The number of nucleotides is constrained to four, but the combinatorics permits an enormous number of possible gene types, far greater than the number realized in any real population, species, or clade. The four-nucleotide limit is not an effective constraint. For cells, tissues, organs, and so on, the number of dimensions of variation is huge, and for many dimensions, like size, variation is not digital and limited as for nucleotides but is more continuous and open-ended. Further, in real populations, most individuals are very similar to each other in many dimensions, as are most species in most higher taxa. In other words, real populations and real taxa are hugely redundant, leaving many avenues for diversification. Consider the enormous amount of variation evident in every species at birth, before the ecological component of selection has had a chance to act. Seemingly, every species has available to it an enormous phenotypic space of the "adjacent possible" (Kauffman 2000). Thus, in the absence of selection, increased differentiation among at least *some* individuals, in *some* dimensions, of *some* parts, is virtually guaranteed in real populations, giving ample scope to the ZFEL to drive diversity upward, despite constraints.

In contrast, death and extinction have the potential to be serious drains on diversity. In extreme cases, reductions in diversity due to extinction have been dramatic. It has been speculated that life on Earth came within a whisker of total extinction during the great Permian mass extinction.

Given present knowledge, it is hard to formulate an expectation, to say whether constraints, selection against diversity, and extinction should be sufficient to block or overwhelm the ZFEL, on average, or not. On the other hand, it does seem clear that over the 3.5-billion-year history of life, diversity has increased, both in terms of number of taxa and in terms of disparity among taxa. Despite some dramatic reversals, the variety of life on Earth has increased—indeed, by some standards, it has exploded—over that time. And this in turn suggests that the ZFEL is stronger than its antagonists, on average. However, we have not yet considered another possibility, that the increase is due not to the ZFEL but to positive selection for diversity. We will discuss this shortly.

The Wind Still Blows. Many will not need reminding, but let us briefly recall that the case for the ZFEL does not hinge on whether forces and constraints are sufficient to block or overwhelm it. The point of the ZFEL is to partition the factors affecting diversity in a new way, one that recognizes a spontaneous tendency for individuals, populations, species, and higher taxa to differentiate and that distinguishes this tendency from other factors. The general formulation of the ZFEL says that an increasing tendency will be present even if constraints and forces are present and powerful. Recall our wind analogy. A shifting wind blowing across a yard scatters concentrations of leaves so that the "diversity" of their locations increases. If an absorbing boundary such as an open garage is present, eventually the leaves encounter the back wall. The spread stops, diversity stops increasing, but the *tendency* to spread is still present. The wind still blows. The same goes for encounters with any other sort of barrier, such as a hedge around the yard. Or suppose that something is reducing the number of leaves, say the homeowner, who has decided to rake up and dispose of as many leaves as she can. Even while she is raking the leaves, even while she is bagging them and carting them away, the wind is still scattering the remainder. The tendency to spread is still present, as long as at least two leaves remain. The wind still blows.

The ZFEL in Evolutionary Theory

The ZFEL is implicit in standard explanations for diversity, especially in the sense of disparity, in certain theoretical contexts. In the original Linnean system, taxa were grouped according to phenotypic similarity (or, more precisely, similarity to an archetype). After Darwin, taxa were understood in terms of common descent, but the Linnean system could be retained because phenotypic similarity corresponds well with propinquity of descent. Linnean genera are more different from each other than species within the genera are from each other, on average, because genera have been diverging longer. The same goes for higher taxa. In standard terms, the reason is that phenotypic divergence is correlated with time. In our terms, the rough consistency between phenotypic and evolutionary taxonomies is an expectation of the ZFEL and evidence for it.

The ZFEL is also implicit in the standard models of phenotype evolution in modern systematics and paleobiology. For example, in maximum-likelihood models for the analysis of character change along phylogenetic trees, character states in a lineage at time $t + 1$ are the result of applying some probability distribution to states at time t.[9] In other words, evolution is a Markov process (Raup et al. 1973; Raup and Gould 1974; Raup 1977). In most models, lineages follow pathways in character-state space that are independent and random with respect to each other, so that the expectation in the absence of constraints or selection is that any two evolving lineages will become ever-more distant from each other. In other words, the null model is ZFEL-driven increase in disparity (e.g., Foote 1996; Gavrilets 1999; Ciampaglio, Kemp, and McShea 2001; Pie and Weitz 2005; Erwin 2007).[10]

Evidence for the ZFEL: Phenotypic Divergence in Macroevolution

The ZFEL is supported by a huge body of evidence. Here we give a subset of that evidence that we think offers especially compelling support for the ZFEL as it applies to diversity—in particular, to diversity in the sense of disparity. (A broader empirical case for the ZFEL will be offered in chapter 5.) The evidence is the divergence of phenotypes on long timescales, especially the diversification of animal life over the past 540 million years, that is, the Phanerozoic Eon, arguably the best known and most widely acknowledged pattern in macroevolution. Our claim will be that the principle underlying what are currently the standard explanations in paleobiology is the ZFEL, although obviously it has not been called by that name in modern discourse. We do not claim to know whether the standard explanations are correct, but we think they probably are, and if so, then the entire pattern of divergence in the Metazoa counts as evidence for the ZFEL.

The basic form of our argument is this: to the extent that different lineages exploit different opportunities, use different resources, adopt different strategies, and so on, changes among lineages will be independent of each other, and therefore the divergence among them is a consequence of the ZFEL. Suppose that one clade of clams evolves a streamlined shape to burrow more effectively into soft mud. And another evolves an acid-producing gland to burrow into hard coral. The resulting phenotypic divergence is the ZFEL. Notice that selection is involved. In this story, it drives adaptive change in both lineages. But the selective forces are somewhat different in each lineage, so that lineages change to some degree randomly with respect to each other. And as a result, they become ever more different from each other, more disparate, with time. Selection is involved, but—and this is critical—not selection *for* divergence. That would not be the ZFEL. (We discuss this non-ZFEL mechanism for producing disparity later.)

Before proceeding, we need to say to something about terminology. We have been using the word "diversity" to encompass both number of taxa and disparity. But the paleobiological literature uses "diversity" primarily to refer to number of taxa, and therefore, to avoid confusion, in the rest of this section we will avoid "diversity" altogether and instead use "disparity" and "number of taxa."

Metazoan Taxa and Disparity over the Phanerozoic. The notion that the number of taxa has risen is uncontroversial. Decades ago, Sepkoski (1978) used data from the paleobiological literature to demonstrate an increasing trend in the number of marine families over the Phanerozoic. His findings have been accepted by paleobiologists generally for about 30 years.[11] But documenting a rise in disparity is more difficult. At present there is no operational set of morphological dimensions that, taken together, would constitute a completely general multidimensional morphospace, a space in which one could plot, for example, a clam, a worm, and fish (Gould 1991). We see no in-principle reason why such a space could not be constructed. And encouragingly, a general morphospace has been developed for metazoan hard parts (Thomas and Reif 1993; Thomas, Shearman, and Stewart 2000). But in the meantime, while awaiting a completely general space, we are forced to rely on proxies. One could argue that number of taxa is a good proxy for disparity, but it is known that number of taxa and disparity can and do change somewhat independently, at least within certain clades (Foote 1996).

In any case, there is evidence that is more direct. Novack-Gottshall (2007) has developed a "theoretical ecospace" framework in which organisms are classified according to 27 ecological characters, including the resources they use, the means by which they acquire or defend those resources, body size, physiology, reproductive mode, and others. In his framework, the character states for each taxon define its "life habit," and the "ecological disparity" of a biota is the variety of different life habits occupied by all of the taxa it contains. Comparing Paleozoic and modern marine biotas, Novack-Gottshall found that ecological disparity was significantly greater in the modern ones. If we can assume that the life habits of an organism are reflected fairly directly in its phenotype, then phenotypic disparity has also increased.

Bambach et al. (2007; Bush, Bambach, and Daley 2007) also have a theoretical framework for classifying marine taxa. The space is defined by three axes representing feeding mode, physical location with respect to the sediment-water interface, and degree of motility. Then, each axis is subdivided into six categories. For example, feeding mode breaks down into suspension feeding, grazing, etc. Thus, the space allows classification of organisms into 216 theoretically possible "modes of life." Bambach et al. found that the portion of the space occupied by marine animals with a good fossil record rose from only 2 modes of life present just before the Paleozoic, to 30 in the early Paleozoic, to 62 to-day.[12] Again, if we can assume that ecological differences are reflected in the phenotype, then Bambach et al. have documented an increase in disparity.[13]

The ZFEL and the Rise of Macroevolutionary Disparity. The literature offers various explanations for this rise in disparity, but most have a common theme: the expansion of life into new ecospace. In general terms, Novack-Gottshall and Bambach et al. argue that new modes of life arose from the use of new resources or the exploration of previously underexploited ones, in a changing environment. One specific proposal has been that the invasion of deep sediments and a rise in carnivory drove much of the expansion. Paleozoic animals were overwhelmingly epifaunal (living above the sediment-water interface) and shallowly infaunal (living just below). But Cenozoic animals also occupy the deeper infaunal realm (Bambach 1983; Droser and Bottjer 1989). The argument is that this invasion of infaunal habitats was driven by predation, with prey animals finding refuge in the deep sediment from mobile Cenozoic predators (Vermeij 1977, 1987; Bush, Bambach, and Daley 2007; Novack-Gottshall 2007).

The possibility has also been raised that post-Paleozoic divergence was driven by the increasing provinciality, or fragmentation, of the world's marine habitats caused by the breakup of Pangaea (Valentine, Foin, and Peart 1978; cf. Miller et al., in press). Finally, Vermeij (1995, 1987) has argued that the major bursts of innovation and new taxa are linked to influxes of nutrients from submarine volcanism, and these new resources facilitated the breaking of ecological constraints, leading to higher rates of adaptation. For example, a snail supplied with more abundant resources needs to forage less, reducing the adaptive constraint imposed by predators and allowing the snail to devote fewer resources to its shell and more to higher metabolism, more elaborate feeding structures, and so on. A resource-rich world is more permissive, with more opportunities for innovation, Vermeij argues.

All of these are the ZFEL. Each species responds to the pressures of epifaunal existence and to the opportunities of infaunal living in a way that is to some degree unique, leading to divergence. The isolation afforded by continental separation leads to independent adaptation to unique local environments, and the result is divergence. In Vermeij's mechanism, constraints and opportunities for innovation will be somewhat different in each lineage, with the result that each changes to some degree independently. The common theme is adaptation in individual lineages to exploit new opportunities, occurring to some degree differently in each species and producing an expansion into unoccupied ecospace. This is the ZFEL.

The ZFEL also seems to underlie the smaller-scale macroevolutionary expansions of the Metazoa. Knoll and Carroll (1999) argue that the divergence of the major phyla—molluscs, chordates, arthropods, and so on—over the first 25 million years of the Cambrian Period, the Cambrian explosion, was the result of developmental-regulatory innovations in a preexisting suite of genes called the genetic tool kit, a rise in atmospheric oxygen allowing an increase in body size, and an extinction event that removed the suite of Precambrian incumbents (Knoll 2003; Carroll 2005).

Close on the heels of the Cambrian explosion, the Ordovician radiation marked one of the most dramatic expansions in metazoan history. Miller (1997, 2004) lists a number of possible causes, including the release of the genetic potential assembled during the Cambrian explosion in an expansion into still-unoccupied ecospace; a flux of mud into the oceans from newly uplifted continents and the advent of carbonate hardgrounds, which provided a congenial habitat for the animals that dominated the diversification; and tectonic disruption of the sea floor environment, which created new opportunities for allopatric speciation.

The radiation of certain mammalian groups in the Cenozoic, over the past 65 million years, may have been the result of a certain key dental innovation, a molar cusp called the hypocone, which is thought to be an evolutionary precursor to a number of adaptive tooth designs, including the square, high-surface-area molars deployed so effectively by ungulate herbivores in crushing fibrous foods. Jernvall and Hunter (1995) have shown that the origins of the hypocone in certain mammalian groups were associated with increases in the number of taxa within those groups.

The great mass extinctions of the Phanerozoic were all followed by recoveries in the number of taxa and disparity. There has been, in paleobiology, some controversy on such issues as the rate at which recovery proceeds (Jablonski 2005; Erwin 2006) and the properties shared by the taxa that survive and those that contribute to the new fauna (Miller and Foote 2003; Erwin 2001, 2006). But this controversy occurs against a background of general consensus about causes, a consensus that these radiations are the result of a refilling of ecospace left at least partly empty by the preceding extinction events.

It should be obvious how each of these expansions invokes the ZFEL. In each case, change is the result of a new opportunity, genetic or environmental, a novel adaptation or the opening of new ecospace, or both. And that opportunity is exploited to some degree differently by each lineage. The result is divergence. And that is the ZFEL.[14]

The above list of explanations for the Phanerozoic expansion is not exhaustive. Non-ZFEL mechanisms have also been proposed. It has been suggested that the Phanerozoic expansion was the result of an increase in ecological specialization and in the degree to which species can be packed into communities (Valentine 1969, 1980; Bambach 1977; Sepkoski 1988). And so it might be argued that, to the extent that specialization is the result of competition among large suites of species, it is the result of selection for divergence in features shared among those species. And in that case, the resulting divergence would not be a manifestation of the ZFEL. (On the other hand, to the extent that packing is the result of the different lineages taking advantage, to some degree independently, of the adaptive payoffs of specialization or evolving in some unique way to escape competition with species using similar resources, the ZFEL is invoked.) We will say more about selection for divergence later. For now the point is that the ZFEL underlies most of the standard arguments for macroevolutionary divergence. It is a screwdriver—a critical and much-used tool—in the standard explanatory tool kit.

The Largest-Scale Divergence. So far we have been talking about animals. But the ZFEL is also a standard explanation for divergence at larger scales. Knoll and Bambach (2000) argue that the increase in number of taxa over the history of life has occurred in what they call megatra-jectories, six expansions of life in which a new realm of ecospace was occupied. These include (1) the expansion from the first protolife to the last common ancestor of modern organisms, (2–4) the expansions of prokaryotes, single-cell eukaryotes, and multicellular eukaryotes, (5) the invasion of land, and (6) the ecological expansion produced by human intelligence. The argument is that each expansion is achieved by the discovery of novel ways to avoid or overcome ecological or pheno-typic constraints, and in each the expansion is driven by a discovery of new resources or new ways of using existing resources. And that is the ZFEL.

Alternatives: Microevolutionary Selection for Divergence. We have not yet considered the possibility that some non-ZFEL *microevolutionary* mech-anism is at work, with effects that propagate up to higher levels. If it is, our argument that the rise in disparity over the history of life is evidence for the ZFEL falls apart. One candidate mechanism is what is now called reinforcement, or what Darwin may have meant by the phrase "divergence of character," in which selection favors variation that removes a population or species from competition with a recently diverged ancestor (Darwin [1859] 1964; Pfennig and Pfennig 2005; Pfennig, Rice, and Martin 2007). The argument here would be that divergence is the result not of different selective forces operating in independent lineages but of a decidedly non-ZFEL selection *for* divergence. Another non-ZFEL candidate is selection for differentiation arising from the reduced fitness of hybrids (Noor 1995; Noor and Feder 2006). In lineages that are already somewhat divergent, and between which hybrids are inviable, infertile, or otherwise suffer reduced fitness, selection is expected to favor variations that reduce interbreeding. In other words, it favors isolating mechanisms, involving changes in sexual anatomy, behavior, etc., leading to further phenotypic divergence. This too is selection *for* divergence.

However, neither divergence of character nor reduced hybrid fitness is expected to propagate to higher levels. The reason is that selection for escape from competition should *decrease* as divergence increases, and it should cease when niches no longer overlap. Millions of years ago, when the terrestrial lineage that would ultimately lead to bats and the one that would ultimately lead to whales had just diverged, they may well have been under selection for being different from each other. But as they diverged, competition would have rapidly decreased. Competition between them is certainly negligible now. The same goes for selection arising from low hybrid fitness. Neither bats nor whales are now under selection for avoiding breeding with the other. Either mechanism might account for early divergence but neither works for long-term divergence.[15]

A clarification may be needed here. It might seem that if, hypothetically, all evolutionary change were driven by selection for divergence or for reduced hybridization in recently diverged pairs of populations or species, then propagation to higher levels would be inevitable. After all, if the only source of change in a system were selection for microevolutionary divergence, then how could macroevolutionary disparity be the product of anything else? The answer is that in such a system change is still random in the with-respect-to sense in most pairs of species. Suppose that selection favors size reduction in some crab species on account of the advantages of not competing with its larger sister species. And at the same time, selection favors a stronger substrate attachment in some clam species, allowing it to move to higher-energy environments and avoid competition with its more lightly attached sister taxon. It is true that the two crab species are diverging under selection for divergence, but each is nevertheless evolving randomly with respect to both clam species—and with respect to virtually every other species on the planet! Likewise, each clam species is evolving randomly with respect to both crab species, and all other species as well. The point is this: even if all species experienced selection for divergence from a sister taxon, most divergence in the system would still be the result of the ZFEL.[16]

Alternatives: Macroevolutionary Selection for Divergence. We are not done. The ZFEL would also be unnecessary if selection could be shown to have favored the avoidance of competition at the level of higher taxa. Instead of selection for non-overlap of niches in recently diverged species, this would involve selection for non-overlap of "adaptive zones" (Simpson 1953; Van Valen 1971) among clades. Or higher-level selection might favor clades with species that are more different from each other, say, if such clades were more likely to survive on long timescales. Clades with greater disparity might be more likely to survive mass extinction events, for example. The problem with these mechanisms is that both invoke higher-level selection, which most evolutionists today are skeptical of. Selection at the level of the totality of life on Earth would require interplanetary competition (differential deaths and/or reproduction) of planet-wide biotas, which is far-fetched. For adaptive-zone-overlap avoidance and clade selection, the objection has to do with the general requirements of selection. Species and clades reproduce, and when they do they exhibit variation and heredity, but selection also requires interaction with an environment. And species and clades do not seem to be good interactors, at least not in the same sense that organisms are (Damuth 1985).[17] Now these arguments could be wrong. Gould (2002) has argued at length for the importance of higher-level selection. And Jablonski (2008) has reviewed a number of empirical studies of species-level properties that seem to be favored by selection,[18] as well as some tantalizing data supporting clade-level selection operating during mass extinctions (Jablonski 2005).[19] We make no judgment here except to say that, as explanations for macroevolutionary disparity, these non-ZFEL explanations are not the standard ones.

Alternatives: Mechanisms for Increasing Genetic Variation and Evolvability. The generation and maintenance of genetic variation within populations are well studied (Hedrick, Ginevan, and Ewing 1976; Hedrick 1986, 2006). For instance, heterozygote superiority can maintain genetic polymorphisms at a given locus. But the idea that this could scale up to explain macroevolutionary disparity is not at all plausible. However, there are other genetic mechanisms that might be. We consider three such possibilities: sex, alternative splicing, and modularity.

The maintenance of sex is favored over asexual reproduction when selective environments are heterogeneous over time and/or space (Antonovics, Ellstrand, and Brandon 1988). Environmental heterogeneity could also have been responsible for the evolutionary origins of sex, but there are other possibilities (Bernstein, Byers, and Michod 1981; Bernstein et al. 1985). But neither maintenance nor origins are relevant to our concerns. Here we want to know whether it is at all plausible that the variation produced by sexual reproduction is responsible for macroevolutionary disparity. Sex has two effects that are pertinent. First, it mixes preexisting variation within populations. Second, sexual lineages may be more evolvable than asexual ones. As for the first, here we repeat what we said above about microevolutionary selection for divergence. There is no plausible way it could propagate upward, no way that the mere mixing of preexisting within-population variants could account for a significant component of macroevolutionary disparity. The second effect, increase in a lineage's evolvability, we discuss below.

Recent work on posttranslational mechanisms suggests that mechanisms for alternative mRNA splicing may be a route to disparity. Small genetic changes, and even environmental changes, can produce mRNA reorganizations leading to novel proteins with novel functions. Alternative splicing was undoubtedly favored on account of the flexibility and fast response time it offers during the lifetime of the organism. In other words, it was favored by individual selection. But it may also have been favored for the fast and flexible response it allows in evolution, its ability to generate new taxa and disparity. In that case, alternative splicing may have been favored by lineage selection for evolvability.

Recent work on modularity of genotype-phenotype mapping has also focused on evolvability (Wagner and Altenberg 1996). The idea is that, with a modular mapping between genotype and phenotype, selection can be much more effective in molding phenotypes adaptively to meet environmental demands. For instance, in mammals forelimbs and hindlimbs can evolve independently, as evidenced by bat wings, whale flippers, and human arms. With a genotype-phenotype mapping that disallowed the separation of forelimb and hindlimb development, these evolutionary outcomes would not have been possible. But when the mapping is modular, great diversity is not only possible but expected.

Interesting as all this work is, it can explain macroevolutionary disparity only if population-level disparity scales upward to the highest phylogenetic levels. It could, in principle, but there are some gaps to be crossed, namely those arising from hierarchical structure. Disparity at one hierarchical level need not propagate to higher levels. A species composed of disparate populations may itself not be especially disparate. A phylum consisting of disparate classes need not itself be very disparate.[20] Thus, for the disparity produced by genetic mechanisms to propagate up to explain disparity in the Metazoa, a number of hierarchical divides would need to be crossed. And we have no empirical reason to think that they have been, or theoretical reason to think that it is likely.

But there is another, deeper, argument to be made here. Consider how these mechanisms for evolvability are supposed to work. Sex, alternative splicing, and modularity are said to be favored at the lineage level on account of their ability to generate disparate phenotypes on which selection can act. In other words, selection works by favoring lineages that are more evolutionarily responsive to environmental changes, those that can be modified independently from other lineages, freed from the historical constraints that otherwise limit morphological evolution. But recall that the independent movement of different lineages through morphospace just *is* the ZFEL. In other words, selection for evolvability *releases* the ZFEL from historical constraints, *allowing* it to generate diversity. It is not an alternative route to diversity.

Other Forces. The possibility remains that macroevolutionary disparity is a by-product of some other force. But this seems unlikely. The rise in disparity has been so common in evolution that the something-else would have to be pervasive, acting strongly across all or most clades. What could this pervasive something-else be? In the absence of an answer, the argument that the ZFEL is responsible seems to us compelling.

Conclusion regarding Macroevolutionary Divergence. What we have offered here is a standard sort of scientific argument for a theoretical point of view. We have invoked a huge body of existing data, encompassing the entire divergence history of the Metazoa, and we have argued that it is best explained by the ZFEL. And further, we have argued that the standard explanations in the field are, implicitly, the ZFEL.

Diversity and the Wind

The ZFEL requires us to rethink the way we view diversification. Whenever diversity is stable or decreasing, the ZFEL says that some other factor—such as loss due to drift into an absorbing barrier or stabilizing selection—needs to be invoked to explain it. And whenever diversity increases, it says that the ZFEL must be at least a contributing factor. Regarding the overall increase in phenotypic diversity in macroevolution, we can say only that the standard account is equivalent to the ZFEL account, and therefore, if the standard account is right, the rise of diversity counts as powerful evidence for the ZFEL.

In closing, let us recall again that the ZFEL tendency, whether augmented or opposed, is always present. In a yard full of leaves, a tendency to scatter is present, even if forces act. The yard owner might assist the dispersing tendency of the wind by raking the leaves outward toward the neighbors' yards. Or she might struggle against the wind, raking them back toward the center of the yard. Whatever she does, whichever way forces act, the tendency to scatter is still present. The wind still blows.

4

Complexity

The complexity of an organism is the amount of differentiation among its parts or, where variation is discontinuous, the number of part types (McShea 1992, 1996). This is the understanding of complexity we adopted in chapter 1. It is decidedly not the colloquial one. In "street" usage, a thing is complex not only if it has many part types but also if it is capable in some way, if its design is impressive and perhaps puzzling, if it is sophisticated, and much more. Our definition is narrower. It is what we are calling "pure complexity," although we will often use just the word "complexity," alone and unmodified, adding the word "colloquial" when we need to refer to the common notion.

Some will be puzzled and perhaps put off by our appropriation of the word "complexity" in this way, even modified by the word "pure." So we begin this chapter by elaborating on our understanding of pure complexity, explaining our word choice and trying to answer objections. We go on at some length not only because of the importance of pure complexity for the ZFEL but because we think that colloquial complexity has been the source of much trouble in biology, that it has been responsible for the paucity of serious empirical treatments of complexity in the biological literature and for an apparent paradox.

(Nevertheless, we do acknowledge that some notion akin to colloquial complexity may be useful in biology. See chapter 7.) We then explain our usage of another term, "parts," which is central to the ZFEL but at present has no generally accepted technical meaning in biology.

Next, turning to the ZFEL, we explain how it applies to complexity and discuss the various forces and constraints that might augment, oppose, and limit it in evolution. The application to complexity precisely parallels that for diversity. This is natural because, in our conceptual scheme, complexity just *is* diversity, at a higher level of organization (chapter 1). The complexity of an organism is just the diversity of parts within it. Finally, as we did for diversity, we present an important piece of the empirical case for the ZFEL.

Pure Complexity

The vertebral column of a typical fish is simple in that the vertebrae are all similar to each other, from one end of the column to the other. A fish column has, roughly speaking, one part type. In a mammalian column, however, we can recognize five vertebral types (cervical, thoracic, lumbar, sacral, and caudal). The mammalian column has more part types, so it is more complex. Or one could decline to distinguish types—since in fact vertebral types do intergrade—and say that the mammalian column is more complex because the vertebrae are more different from each other, on average, than in a fish. In other words, the variance among parts is greater in a mammalian column.

Notice that, as with the term "diversity," "complexity" in our usage has two senses, discrete and continuous, "part types" and "differentiation among parts" (paralleling "number of taxa" and "disparity among taxa" for diversity). Both are aspects of the same concept, variance, and in what follows, we switch somewhat casually from one to the other, employing types or differentiation as needed. This is allowed because the ZFEL applies to both (although, as we will point out shortly, the special formulation applies without qualification only to the continuous sense).

Notice too that our understanding of degree of differentiation does not take into account absolute number of parts. A vertebral column with 40 vertebrae is no more complex than one with 30, assuming both have the same number of types or are equally differentiated. This indifference to absolute number of parts makes sense, in this context, because the ZFEL predicts an increase only in types and differentiation, not in numbers. On the other hand, numbers can, under certain circumstances, act to constrain complexity in our sense (see below).[1]

Complexity as a Level-Relative Concept. Different complexity values will typically be found at different levels of organization in the same object. Consider a fish, not just its vertebral column but the whole animal. At the cell level, its complexity is the number of cell types, about 120 for the few species in which counts have been attempted. But at a higher level, the level of tissues and organs, it has a different value, about 90. And at a lower level, say the atomic level, the fish's complexity is the number of different types of atom it contains, the half dozen or so elements present in any appreciable quantity.[2] The point is that the different complexity values at different levels present no contradiction. Pure complexity, like diversity, is simply a level-relative concept (McShea 1996).

An important clarification: pure complexity is level relative but is not a function of number of levels. In contrast, colloquial complexity has a component corresponding to what might be called hierarchical depth, or the degree to which a system is physically nested, the number of levels of parts within wholes it contains. By this criterion, a multicellular eukaryote contains more levels than a solitary prokaryotic cell; a society of multicellulars, more levels than a solitary multicellular individual. In evolutionary studies, there has been considerable interest recently in the rise of number of levels of selection (Maynard Smith and Szathmáry 1995; McShea 2001; Marcot and McShea 2007). Hierarchical depth is an important feature of organisms. And it can be formulated as a pure concept, divorced from any notion of fitness and function, based solely on number of levels of nestedness (McShea 2001). And in that case, hierarchical depth would refer to what has been called the "vertical" aspect of pure complexity (Sterelny 1999). But the ZFEL makes no prediction regarding hierarchical depth. The ZFEL is relevant, so far as we know, only to the "horizontal" aspect of pure complexity, number of part types at the same level. This is not to deny that there might be interesting relationships among horizontal complexity values across levels (McShea 2002). Rather, it is to say that the ZFEL prediction is always specific to a particular hierarchical level. More precisely, it predicts increase (in the absence of forces and constraints) at *every* single hierarchical level independently, regardless of how many there are.[3]

Pure and Colloquial. This view of complexity as part types or differentiation is fairly intuitive. A calculator with many part types seems more complex than a sundial with one or two. A formal dinner place setting with seven different utensils seems more complex than an informal one with three. A contract with more terms would generally be thought of as more complex than one with fewer. And where differences among parts are continuous rather than discrete, intuitively complexity increases with degree of differentiation among them. A bouquet of flowers with both red roses and pink chrysanthemums is more complex, at ordinary scales of observation, than one with only red and pink roses. Roses and chrysanthemums are more different from each other than are red and pink roses (since both are roses). A Swiss Army knife with three different kinds of cutting tool—say a knife, a saw, and a can-opening blade—is more complex than one that only has three different sizes of knife. All of these are cutting tools of some kind, but the knife, saw, and can opener are more different from each other than are the three knives.[4] Likewise, bipeds typically have a more complex limb structure than quadrupeds. The fore- and hindlimbs of a kangaroo, a bird, or a human are more different from each other than are those of a cat or a cow.

But pure complexity is also nonintuitive, insofar as it fails to align with the colloquial usage. As discussed in chapter 1, colloquial complexity is a function not only of differentiation or number of part types but also of functional capability, degree of organization, difficulty of manufacture, degree of sophistication, and much more. For example, a human brain is considered complex in the colloquial sense not only because of the number of cell types it has or the number of folds in it but also because of what it can *do*. In contrast, complexity in the sense of part types and parts differentiation does not depend on ability to *do* anything. Thus, a live organism and a dead one are equally complex in the pure sense if they have the same part types. A laptop computer may be less complex in the pure sense than one that has been dropped from a high place, if the impact shatters its internal mechanisms into hundreds of pieces, each a new part type. A human with wisdom teeth or an appendix is more complex than one without them, regardless of whether these parts are useful, neutral, or hazardous to their owner. All of this will sound odd to those steeped in the colloquial meaning. To them, pure complexity will seem not only unintuitive but thin, colorless, even dull.

Against Colloquial Complexity. Not every sexy-sounding idea is useful in science. And colloquial complexity may be one of those ideas that is not. One big problem with it is that it seems not to have any consistent meaning. Colloquially we might say that the blade of a knife is simple because it consists of a single part, but we might also say it is complex if the manufacturing process has many expensive or difficult steps. Colloquially we might say that a loan agreement is complex if we do not understand it, but we might call a recipe complex on account of the number of steps or the coordination it requires, regardless of whether we understand it. Worse, colloquial complexity often has an evaluative component. In calling a machine complex, we sometimes mean that it is *better* in some instrumental sense than simpler ones. And when applied to organisms, complexity sometimes has a moral dimension of a sort that is simply forbidden in science. When humans are called complex, the implication is sometimes that we are better than other species in some absolute, noninstrumental sense.

Given all of the different and apparently incommensurate aspects of colloquial complexity, it is reasonable to wonder whether an operational definition of colloquial complexity is possible, whether there is any way to make it measurable and useful for research. Indeed, one is left wondering whether colloquial complexity has any real use in biology beyond expressing, perhaps with useful ambiguity, the feeling of awe that organisms often (rightly) inspire.

We will go further. In biology, colloquial complexity has been a mischievous elf, stirring up trouble where none existed, leading astray otherwise-interesting research. This emerges clearly in the history of the infamous C-value paradox. Decades ago, the very reasonable hypothesis was raised that number of genes in an organism might be correlated with complexity of form. The thinking was that a more complex organism ought to require more instructions for its development, and therefore more genes. How to test this? The technology of the 1950s and 1960s could not provide direct gene counts, so amount of DNA was used as a proxy. Complexity was interpreted in the colloquial sense and therefore could not be measured, so an ordinal ranking was substituted. Humans were taken to be the most complex, followed by other mammals, then amphibians, arthropods, and so on. Armed with these two proxies, testing became possible. But early results did not support the hypothesis. The correlation between amount of DNA and the proxy ordination was poor. Notably, some low-ranking organisms, including some amphibians, have huge amounts of DNA compared with high-ranking ones like humans—hence the so-called C-value paradox. Today we can more directly count genes (although the definition of "gene" is becoming increasingly ambiguous). Humans seem to have on the order of 25,000 genes, while fruit flies have about 13,000, and the nematode *Caenorhabditis elegans* has about 19,000. Humans do seem to have more genes than these "simpler" organisms, but not many more, considering how much more complex—in the colloquial sense—we imagine ourselves to be. And so the apparent paradox persists.

From our perspective, however, the gene counts from fruit flies, nematodes, and humans do not present even a puzzle, much less a paradox.

Colloquial complexity has never been operationalized. So, if a good correlation had been found between colloquial complexity and gene number, it would have meant virtually nothing. It would have meant that something-we-do-not-really-understand is correlated with gene number. In our view, biology simply took a wrong turn here, ignoring the usual scientific standards and invoking this poorly defined concept. And it is pretty obvious why it did so. The human-amphibian-arthropod ordination is just the secretly much beloved but officially reviled Great Chain of Being (Sealfon 2008), and complexity in its colloquial sense has become a convenient code word for it. This is not to say there is nothing worth investigating here. The relationship between morphological complexity and number of genes remains potentially an interesting one, if only we could operationalize complexity independently of the Great Chain. Pure complexity is one way to do that.

The Virtues of Pure Complexity. Pure complexity has been applied successfully in evolutionary studies in recent years. Marcus (2005) studied pure complexity in bacteria (and interestingly found a statistically significant correlation with gene number). And studies have been done of evolutionary change in differentiation among, for example, vertebrae in vertebral columns (McShea 1993; Buchholtz and Wolkovich 2005), number of cell types in animals (Valentine et al. 1994), number of different curvilinear components in the sutures marking septa in ammonoids (Saunders, Work, and Nikolaeva 1999), number of types of bony elements in the skulls of mammal-like reptiles (Sidor 2001), and number of limb-pair types in arthropods (Adamowicz, Purvis, and Wills 2008)—all offered as investigations of complexity in what we are calling the pure sense.

Probably what will seem most strange about pure complexity is its icy indifference to function. But we consider this a virtue, for the two reasons mentioned in chapter 1. One is that it enables us to ask "how many" questions. Now, for some, the ability of organisms to function, to perform tasks, behaviorally or physiologically, is what biology is all and only about. Was it not Darwin's explanation of function that transformed biology into a science? We agree that function is important. But function is not the only interesting issue in biology. And pure complexity is about something different. Consider once again the concept of diversity. One measure of diversity is the number of different species in, say, an ecosystem. But in measuring diversity, we do not ordinarily ask about the *function* of those species, what they can do *for* an ecosystem, or the level of sophistication with which they do it. Or if we do ask these things, it is clear that we are asking questions that are separate from the pure diversity question. We may think of diversity as a possible correlate or predictor of ecological structure, stability, or energy usage, but these are things we hope to discover, not concepts built into the definition of diversity. Diversity is about variety among individuals or taxa, regardless of "ecological function." Likewise, pure complexity is about variety among parts within an organism, regardless of how they function. And if we do ask about their function, it is clear we are asking a different question. For complexity, the colloquial usage blurs the "how many" and "how capable" questions together. Pure complexity is only about "how many," and thereby keeps them separate.

A second advantage of the divorce from function is that it enables one to actually investigate the relationship between part types and function, or part types and natural selection. Are organisms with more part types favored, on average? Or equivalently, since function arises by natural selection, we could ask whether organisms with more part types are more functional, more fit. Questions along these lines have been posed, especially in paleobiology. For example, there have been studies of the relationship between organismal complexity and extinction susceptibility—a proxy for fitness—in higher taxa (e.g., Schopf et al. 1975). As these studies have implicitly recognized, finding a relationship between complexity and fitness makes sense only if one starts with conceptually nonoverlapping notions of complexity and fitness. To put it more baldly, it is virtually impossible to investigate the relationship between complexity and any notion of functionality, including fitness, if one's concept of complexity has some notion of function built into it, as colloquial complexity does. *This point cannot be overemphasized. It is a problem that we think any scientist addressing the relationship between function and a functionally laden notion of complexity needs to answer.* Of course, using pure complexity, or any function-free notion of complexity, there is no problem. Thus, our decision in this book to isolate complexity from function is not merely a consequence of our interest in the ZFEL. Such a separation is a necessary first step for anyone interested in the relationship between the two.

Measures of Pure Complexity. For present purposes, when we use "complexity" in the generic sense, without reference to any particular biological data set, we intend it to include both the discrete (number of part types) and the continuous (degree of differentiation) sense. And, as with diversity, in particular contexts, we can operationalize complexity ad hoc, in whatever way is most suited to the data at hand, because the

ZFEL is applicable to all intuitively reasonable, pure measures. Number of part types has proven an apt measure in many contexts, such as Valentine, Collins, and Meyer's (1994) cell-type study and a study of numbers of part types within cells by one of us (McShea 2002). Information-theoretic measures, such as the Brillouin equation, have also been used, as in Cisne's (1974) study of complexity in arthropods. Continuous measures of complexity that have been used include the variance among parts and the range of variation (McShea 1992; Buchholtz and Wolkovich 2005), paralleling the various measures of disparity (see chapter 3). All are apt in that all capture some aspect of degree of differentiation among parts, and the ZFEL predicts that all will increase, in the absence of selection and constraints.

We need to be careful here. The literature contains quite a number of other usages of the term "complexity" and other operationalizations of it. Many are not directly concerned with complexity in our pure sense and therefore are not relevant here. But others do have some relationship to pure complexity, and we want to say that we do not intend to denigrate them by omitting them from the discussion. In particular, some from physics, mathematics, and computer science do not try to capture all aspects of the fraught colloquial notion but instead seek to measure certain aspects of it, including pure complexity. (See a review by Adami 2002.) Indeed, some are fully compatible with complexity in the pure sense.[5] One reason for not saying more about them here is that it is not obvious how to apply the ZFEL to them. Another is that deploying them in biology—to measure the complexity of real organisms, at a variety of hierarchical levels (as opposed to, say, just the molecular level)—is also not straightforward. In contrast, complexity in the sense of part types and differentiation is straightforwardly applicable at every level and, further, has a track record of successful application.

As for diversity, we need to point out that complexity in the continuous sense is more general, and that the special formulation of the ZFEL applies without qualification only to complexity in this sense. And the reason is the same. If number of parts happens not to be increasing, number of types can increase only until each part is a unique type. In other words, discrete measures are—on account of their built-in constraints—limited in their ability to detect increasing differentiation, which continues beyond the point at which all parts are unique. Of course, this is a constraint on the ZFEL only as the limiting case is approached. Thus, while we include both the discrete and continuous sense in our understanding of complexity, when we use the term generically, we do so with the implicit and sometimes explicit caveat that the special formulation of the ZFEL applies only so long as the constraints imposed by the discrete measures are distant, that is, so long as they are not effective.

Why Choose the Word "Complexity?" It may seem that we have done some violence to the colloquial meaning by appropriating the word "complexity" in this way. Even if it were granted that part-types-and-differentiation is one aspect of what makes organisms so complex in the colloquial sense, did we have to use the word "complexity"? Could we not have picked a word like "complicatedness" instead? Or could we not have invented a new word, like "somethingelseness" or "differentiatudinosity"? We opted against this, first because the word "complexity" presently has considerable heft in science despite its vagueness (or perhaps *because of* its vagueness), and it has a strong grip on the imagination. And we think the notion of part types and differentiation is important enough in biology to be worthy of a word with that heft and power.

Second, pure complexity may be a useful stepping stone on the way to understanding colloquial complexity. We think the present colloquial notion is compound and soaked in ambiguity, making direct application of it in science a quixotic exercise. (Recall the C-value episode.) But we acknowledge that that could change. Colloquial complexity, or something like it, may yet turn out to be scientifically useful. (Indeed, in chapter 7, we partly revive it in order to say something about its relationship in evolution to pure complexity.) And one way to begin to rehabilitate colloquial complexity may be to tackle its components one at a time, starting with part types and differentiation. In other words, the part-types-and-differentiation notion of complexity can be seen as a first step toward the colloquial one. And in that case, including the word "complexity" in the label—pure complexity—advertises its critical role in seeking to understand what could one day be the grander notion.

An Invitation. In adopting this notion of pure complexity, we are aware that we are making an unorthodox move, and we must acknowledge that many readers will not be enthusiastic about making it along with us, despite our arguments. And to them we offer not more arguments but an invitation, to accept our usage temporarily, to play along, to see where this goes.

The ZFEL for Complexity

For diversity, we began with a population of identical organisms. Here we begin with a single organism—or, if sexual, a pair—consisting of a set of identical parts. Imagine a wormlike animal consisting of a series of identical segments. In the absence of natural selection or any other force or constraint, mutation will cause the segments of the offspring to differ from each other. And in each later generation, differences among the segments will accumulate, causing them to become ever more different from each other. More precisely, within an individual, for any set of two or more parts sharing some dimension, those parts will have some distribution in that dimension.[6] The ZFEL says that in individuals in later generations, in the absence of selection and constraints, the variance of that distribution will tend to be higher.[7] For continuous characters, the distribution is expected to spread, with the upper and lower tails diffusing up and down respectively. For discrete characters, the increase may involve the diffusion and redistribution of parts among existing states, or it may occur via the addition of novel extreme states, extending the number and range of states realized, or both. In other words, in the absence of constraints and selection, the ZFEL predicts an increase in pure complexity. That is the special formulation of the ZFEL (see chapter 1). We could, alternatively, invoke its general formulation, in which case we would say that there is a *tendency* for complexity to increase, one that acts whether or not constraints or selection are present.[8]

As for diversity, the ZFEL rationale is simply that random variation occurs, affecting different parts differently and therefore making them ever more different from each other. The picket fence becomes more complex.

Of course, as in the particle model in chapter 2, the expectation of increase is probabilistic. By chance, complexity *can* decrease, in that parts can vary in such a way as to become more similar to each other. A number of long segments could by chance get shorter at the same time that a number of short segments by chance get longer. Of course, this would be less likely in a more realistic model with higher dimensionality. Real parts have many dimensions, not just length but also width, height, mass, energy consumption, growth rate, and so on. And the few dimensions in which parts by chance become simpler will typically be overwhelmed by the greater number in which they become more complex. Thus, complexity in any one dimension is expected to increase, but overall complexity is even more certain to increase.

Organisms as Redundant Systems. The opposite of complexity is redundancy, similarity among parts. To understand the ZFEL prediction, it helps to recognize just how redundant organisms are. One liver cell is very like another liver cell, left sides are like right sides, hands are like feet, and so on. Even very different parts reveal some degree of redundancy. Muscles are like tendons in some respects. A stomach is a little like an intestine. The ZFEL claim is simply that random variation will tend to destroy redundancy, in other words, to create complexity.

Randomness. The ZFEL is driven by random variation. As with diversity, randomness can be understood in either of two senses. Random variation in parts can be understood as morphological drift, perhaps underlain by drift at the genetic level. Or it could be understood to result from deterministic forces acting on each part independently. In this second case, parts change deterministically but randomly "with respect to each other" (see chapter 2). It could even be the case that each part is tightly controlled by natural selection. Consider two structures in an organism, both evolving under tight selective control for different functions, say the limbs of a vertebrate evolving toward bipedal running. The hindlimbs might be selected for balance and support while the forelimbs are being selected for grasping. In our hierarchical perspective, they may be said to be evolving *randomly with respect to each other*, at least to some extent. From the ZFEL perspective, this situation is indistinguishable from drift, and the ZFEL makes the same prediction. It says that the limb pairs will tend to become ever more different from each other, and the organism as a whole will tend to become more complex. In sum, to the extent that selective forces are independent, they can be said to be random with respect to each other. For the ZFEL, it does not matter whether the source of change is a true sampling process, such as drift, or randomness-with-respect-to, that is, causal independence.

Consider another example. In bilateral structures in animals, there are powerful selective forces and developmental constraints enforcing similarity between left and right. That explains why left and right limbs, for example, tend to be similar. On the other hand, in some cases, constraints are relaxed or broken, and selection acts on each limb independently, as in lobsters, where it has likely been at work, favoring hypertrophy of one claw for one purpose and a lesser size for the other for another purpose (Palmer 2005). Here, to the extent that the bilateral constraint and selection for similarity are absent, and to the extent that these changes are driven by different selection pressures acting on each claw, the changes can be said to be random with respect to each other.[9] And the resulting increase in complexity is the ZFEL in action.

Parts. We have been using the word "part" as if it were unproblematic. But the notion of "part types" is not in the conceptual tool kit of most working biologists, and some may be uncomfortable with it. Parts in organisms are not always cleanly identifiable like those in machines. Consider whether, in a human, a heart is a part, distinct from the rest of the circulatory system. Or whether a hand is a part at the same hierarchical level as the rest of the arm, or whether it is a part at some smaller scale, a kind of "subpart" of the arm. Common intuitions do not offer simple answers, and the accidents of anatomical terminology generate even more confusion.

Nevertheless, an intuitively reasonable a priori understanding of parts can be devised and used to identify parts in real organisms. One of us has done so (McShea and Venit 2001; McShea 2002), based on treatments of parts and hierarchy by Simon (1969), Wimsatt (1974), and Salthe (1985). A part is a set of entities that are well connected to each other and relatively isolated from other entities outside the set. In other words, parts are internally integrated and externally isolated units. We will not explain in detail here. For present purposes, it is enough to say that boundaries are important indicators of isolation—walls, membranes, spaces, and such—so that objects separated by boundaries, like the organs in a vertebrate, are distinct parts. But there are other indicators of isolation, such as discontinuities or changes in composition and shape. For example, a pseudopod of an amoeba is marked by a discontinuity in shape—a difference in shape between it and the rest of the cell membrane—and is therefore a part. It is a transient part, of course, but no less a part while it exists. Further, beyond "object parts" like these, there are parts in which the connections are processes or interactions, so that the entities constituting the part may be quite dispersed. For example, the connected group of nerve cells, muscles, bones, and so on that produce a behavior is a part, that is, a "behavior part." And the set of molecules that interact in a physiological cycle is a "physiology part." This scheme also has a hierarchical aspect. If a part is a set of components that is connected internally and isolated externally, then a "subpart" is a connected-isolated set lying topologically within that part.

This understanding agrees with intuition in clear-cut cases, like the parts in a machine and most cells in multicellular organisms, but it also allows for an unconventional possibility, the idea that "partness" might be a quality that comes in degrees. The heart may be somewhat a part, with its degree of partness dependent on its compositional and shape differences from the adjacent vessels. Hierarchy too becomes a matter of degree, so that some parts may lie somewhat at one level and somewhat at another (Wimsatt 1974). This notion of degree of partness is not yet operational, however, and therefore, in our thinking about the ZFEL, we avoid hard cases like hearts and hands. Where an operational definition is critical, we stick with the paradigmatic parts, the discrete or quasi-discrete objects within organisms, like segments, organs, cells, and so on (all the while aware that even these are parts only most of the time, at certain stages in development, and, even then, imperfectly).

What will strike many biologists immediately about our understanding of parts is the absence of any notion of function. Is it really possible to talk about parts in organisms in this way? In biology, at least, the link between function and parts will seem inescapable to many. In our view, there is a connection, namely that there are good reasons to think that parts are very much the result of natural selection favoring internal integration and some degree of external isolation in functional units (McShea 2000). Parts arise in organisms as functional modules, of a sort.[10] But this is different from a claim that parts are *defined* functionally. And in fact, they need not be. Consider that early anatomists were able to identify parts in organisms without knowing their function. Parts are defined and can be identified by boundaries, by isolation.[11]

Our notion of parts has a number of virtues. First, assessments of complexity—such as counts of part types—in real organisms can in many cases be made objectively. Independent biologists should identify roughly the same parts. Also, this notion of parts has been applied successfully in certain empirical studies, for example, a study comparing the complexity of cells in protists and in multicellular organisms (McShea 2002). That study found that protists are more complex than individual cells in plants and animals. What is important in the present context is that the difference was statistically significant, which indicates that the "parts" being counted had at least some reality. In any case, for present purposes, we do not need to defend any particular understanding of parts. All that we require is that there be *some* objective and function-free notion of parts that can be meaningfully applied to organisms.[12]

Heredity and Reproduction. As in the discussion of diversity, we take it to be obvious that organisms reproduce and that their parts and the properties of those parts are to some degree heritable.

Scope of the ZFEL for Complexity

The ZFEL's scope in biology is huge. It applies to any set of homologous parts. Segments of a worm will tend to become ever more different from each other. Ulnas will tend to become ever more different from tibias. But it also applies to any set of parts sharing a common dimension, homologous or not. Heads should become ever more different from hearts. A head has a length, for example, and so does a heart. In the absence of selection and constraints, the lengths of both will vary randomly. The ZFEL predicts that whatever the initial difference between them in length, that difference will tend to become greater rather than less, on average. And this applies also to every other shared dimension, such as the difference in their widths, in the number of cells they contain, in the metabolic rate of those cells, in the *color* of these two organs, and so on. It applies to all sets of parts, no matter how similar or different initially.

And the ZFEL should apply at a wide range of levels of organization, predicting an increase in complexity at the level of macromolecules, organelles, cells, tissues, organs, segments—any level at which parts can be identified.[13] Further, it applies equally to all organisms, from bacterium to blue whale, any evolving organism in which parts can be identified. Finally, the ZFEL applies on all time scales. At the shortest time scale, a single generation, the ZFEL predicts that, in the absence of selection and constraints, offspring will be more complex than their parents, on average. And to the extent the process is cumulative, complexity should tend to increase over many generations, over thousands, millions, and tens of millions of years.

Combining all these domains of application leads to the following claim, which seems to us inescapable, however immodest sounding: *The ZFEL predicts a tendency toward increasing complexity in every set of parts, in every evolving species, on every time scale, over the whole 3.5-billion-year history of life.*

Contrary Tendencies and Constraints

The general formulation of the ZFEL says that complexity has a tendency to increase. But a tendency is not a result. In principle, it could be that constraints block the increasing tendency at every turn. Or it could be that there are contrary tendencies, driving complexity downward. In what follows, we outline some of these constraints and tendencies (although the discussion of the most important candidate for a contrary tendency, natural selection, is saved for the next chapter). The point here is not to argue that constraints and contrary tendencies are minimal, although some clearly are. Nor do we argue that they do not impede the ZFEL tendency. On the contrary, some may entirely dominate it.

Rather, the point is to redraw the conventional picture of the relationships among change, force, and constraint in the evolution of complexity, as we did in the last chapter for diversity. In the conventional view, change is the resultant of forces and constraints. In the redrawn picture, change is the resultant of forces and constraints plus a background increasing tendency, the ZFEL. Or, recalling the analogy with Newtonian mechanics in chapter 1, the ZFEL tendency is like inertia. It is a background condition that is always present and on which forces and constraints are imposed.

Developmental Tendencies. Random variation would certainly raise complexity if organisms were like picket fences and did not reproduce or—allowing that organisms do reproduce—if they copied themselves directly the way DNA or cell membranes do. Almost any variation in the phenotype would have the immediate effect of increasing its complexity. But at least in multicellular organisms, genetic variation is transduced by development. And without knowing how development is structured, there is no reason to think that randomization of the DNA *must* be complexity generating in the phenotype.

There is some reason to think it *might* be, however. Similar developmental processes must underlie the generation of similar parts, and mutations with local effects must inevitably arise, affecting one iteration of the shared generative process differently from another. Similar parts will tend to become different, and dissimilar parts to become more dissimilar. Differentiations of the various developmental modules associated with the *Hox*-gene complexes in arthropods and vertebrates are examples of this.

On the other hand, there is also reason to think that development is structured in such a way as to make spontaneous complexification improbable. It is rarely stated formally but is widely acknowledged informally that the loss of parts in development ought to be easier, or more probable, than gain. The rationale has to do with standard assumptions about development's hierarchical structure (e.g., Riedl 1977; Wimsatt 1986; Arthur 1988). Organisms start simple, as embryos, with few part types, and they become more complex as new part types arise. It follows that developmental processes must be organized at least to some extent in few-to-many causal cascades, in which a small number of early parts give rise to a larger number of later parts. And to the extent that this is true, defects occurring early in development have the potential to eliminate whole suites of parts at a single stroke, indeed, to eliminate all parts lying developmentally downstream (Wimsatt 1986). Given this hierarchical structure, elimination of part types might seem very easy, much easier than the production of new part types, which must—in one way of thinking—require novel genes, novel morphogens, novel induction events, novel regulatory apparatuses, and so on. In other words, the argument goes, the hierarchical structure of development imposes a bias favoring part loss over part gain.

There is considerable evidence for hierarchy. For example, some genes are known to control large suites of other genes and to initiate long hierarchical cascades leading to the development of complex structures, that is, structures containing many part types. And damaging mutations in these genes can lead to the deletion of many parts or, what is the same thing from a complexity point of view, their replacement with redundant copies of other parts. For example, in the *Drosophila* mutant antennapedia, the antennae are lost and replaced by rough copies of legs. The loss of the aristae is a loss of a part type, but the legs are not a new part type, and therefore this is a reduction in complexity.

But it is also recognized that development is imperfectly hierarchical. Defects in early development need not eliminate parts if the defects are compensated for by buffering mechanisms. And defects that cut developmental pathways can also allow the expression of underlying default pathways leading not to part loss but to variation and differentiation of existing parts. Further, while error can produce developmental loss of a part, loss can lead to new tissue contacts, which in turn can lead to novel inductive events, generating new part types and novel modifications of existing parts. For example, Müller and Streicher (1989) have shown that, in the evolution of bird hindlimbs, a developmental event—the spreading of a tendon from the fibula onto the tibia—apparently induced the formation of a novel structure, the syndesmosis tibiofibularis, which later became critical in force transmission between the two bones. The point is this: tissues can respond actively rather than passively to perturbations (Goodwin 1994), producing new structures as well as losses.[14]

Finally, we suggest that the common intuition that development is biased in favor of part-type loss may be partly a consequence of the difficulty we have in imagining how new types could arise. An excellent review of developmental and evolutionary routes to complexity in animal nervous systems by Oakley and Rivera (2008) offers a counterweight to that intuition. Here are three of the routes they identify: (1) copying and divergence, in which an ancestral part is copied and the two copies diverge in structure, as occurred in the evolution of the gene networks of the rod and cone phototransduction cascades (i.e., one part type giving rise to two); (2) splitting and divergence, where an ancestral part splits into two subparts that then diverge, as in the differentiation of the neural crest cells in chordates, which may have occurred from a fissioning of the ancestral neural plate (again, one part type giving rise to two); and (3) copying or splitting followed by fusion, in which two ancestral part types either split or are copied and half of one pair fuses with half of the other, producing a third type. As a possible example of this third route, Oakley and Rivera suggest vertebrate eye development, in which a portion of the neural tissue contacts the overlying ectodermal tissue, inducing the formation of a third tissue type, the eventual cornea. They offer many more examples, spanning a great hierarchical range, from the single gene to the gene network to the tissue and organ level. Our point here is that there are a number of simple developmental routes to new part types, all of them common in evolution.

The Wind Still Blows. The conclusion we intend to be drawn from the discussion above is that the net effect of the structure of development on complexity is unknown.[15] *But even if it were discovered that development imposes a net bias toward loss, such a bias would not challenge the ZFEL.* And the reason is simply that the ZFEL (in its general formulation) points to a tendency, and this tendency acts whether or not some contrary bias is present. Thus, even as some structural bias in development is eliminating part types, the ZFEL tendency is adding new ones. In other words, any bias toward decrease built into development achieves its effect by first overcoming the ZFEL increasing tendency. It is as though some force were removing pickets from the picket fence, perhaps a tendency for the pickets to rot and disintegrate. As this happens, the complexity of the picket fence declines. But this decline is separate from the accumulation of differences among the remaining pickets, which continues unabated at the same time, *tending* to make the fence more complex.[16]

Recall the blowing-leaves analogy from chapter 3. Here we are talking about complexity, not diversity, so the dispersal of leaves is analogous to the increase in complexity of parts in an organism rather than the diversity of organisms. As in the earlier analogy, suppose there is a garage in which leaves accumulate, and this structure is responsible for a decline in the dispersion of the leaves in the yard over time. By analogy, development could be such a structure, which in standard evolutionary thinking could be properly identified as a cause of a decline in complexity. The ZFEL view would not challenge this conclusion, but it would invite a gestalt shift, a change in what is foreground and what is background in our understanding of this situation. In the ZFEL view, the background tendency at work is the wind, tending to disperse the leaves. Then, in the foreground, superimposed on the dispersing tendency, is the structure of development, with properties that cause the leaves to collect together, drawing complexity down. In the ZFEL view, the cause of the decline is a factor promoting decrease superimposed on a background that promotes increase. Thus, we are entitled to say that, even while dispersal decreases, a positive dispersing tendency is present and acts, unabated. The wind still blows.

One might reasonably ask what justifies this move, putting the ZFEL in the explanatory background and development in the foreground. In chapter 6, we discuss the backgrounding of the ZFEL in broader terms. Here we will just make the point that the ZFEL is more general than any particular organization of development. It could be, for example, that the structure of development in most organisms on Earth imposes a bias toward complexity decrease. But the ZFEL imposes a tendency toward increase in *all* organisms, here and elsewhere. The ZFEL is the more fundamental factor, and more fundamental factors belong in the background. For complexity, the ZFEL is where explanation starts. It is the river in which developmental rocks and jetties may be strewn.

Other Constraints. Other than the structure of development, what constraints affect the expression of the ZFEL? Complexity increase is constrained in principle by what might be called the memory of the system, or the amount of storage capacity for differences. In the genome, one limit is the amount of genetic material. Other things being equal, an organism with more genes can store more differences among genes than one with fewer. At higher levels, the memory limits are the number of molecules, multimolecular structures, organelles, and such, and—for organisms with even more levels of organization—number of cells, tissues, and so on. In effect, the limitation is body size. Large organisms can store more differences among parts than small organisms. In a large organism there is simply more material to become differentiated.

In organisms with small and highly streamlined genomes, the in-principle genome-size limitation could be an effective one. But in organisms with genomes rich in duplicated genes, pseudogenes, repeat sequences, and so on, it probably is not. Every bit of genetic redundancy is an opportunity for further differentiation. And at a higher level of organization, body size is unlikely to be a significant limit except for the smallest organisms, given the enormous redundancy of parts within real individuals. In a mouse, for example, complexity at the cell level is limited by the number of cells. A mouse has many millions of cells and therefore could in principle have many millions of cell types, if each cell were a unique type. But in fact a mouse has only a couple of hundred cell types, far short of the theoretical limit. Even organs are not limited. An elephant has about the same number of organ types as a mouse, but in principle, if its organs were mouse sized, it could have a million times more. In other words, the structure of real organisms leaves room for huge increases in complexity. Of course, there are good reasons why a mouse does not have millions of cell types and an elephant does not have millions of different organs, but those have to do with natural selection, not with limits imposed by body size.

Another possible limit is the number of dimensions available to vary and the range of variation possible in each part. Of course, real organisms are spectacularly multidimensional, each part having as many dimensions as it has characteristics (length, width, mass, rigidity, growth rate, color, etc.). Limits do exist, perhaps in most dimensions of variation. But to block an increase in differentiation between any two parts, variation would have to be blocked in all of them.

Complexity and the Wind

We need to make clear once again that our argument for the ZFEL does not hinge on the absence of limits. True, the special formulation does require that constraints be absent or distant. When they are, it predicts that complexity will increase. But when they are not, the special formulation simply does not apply. In contrast, the general formulation of the ZFEL always applies, whether or not limits have been reached. A turbulent wind scatters the leaves on a lawn ever more widely. If there is a hedge around the lawn, dispersion eventually reaches a limit and ceases. But the tendency to disperse continues.

Red Herrings: Novelty and Exaptation

This will be a short section, because novelty and exaptation are not directly relevant to the ZFEL. Novelty, as the term is generally used, refers to adaptive novelty (Moczek 2008), and exaptation is the conversion of a structure with one function into one with a different function. And while the ZFEL is concerned with the origin of new structures and new part types, it is not directly concerned with whether or not they are functional.

Also, and more to the point, the advent of a novelty or an exaptation does not by itself guarantee that the result will be an increase in complexity. Obviously, complexity would increase if a new part type arose or if a modified part were a copy and the original did not change (or if it changed differently). But a novelty could also arise by transformation of a single-copy part or by the fusion of two single-copy parts. In the first case the result would be no change in complexity, because the original part type would have been lost in the transformation, offsetting the gain of the novel type (Oakley and Rivera 2008). In the second case, the result is a decrease in complexity. In sum, knowing that a novelty has arisen or that an exaptation has occurred tells us nothing about complexity.

Evidence for the ZFEL: Pseudogenes

As in the diversity chapter, we close with an example, an instance of the ZFEL that we think both illustrates it and constitutes strong evidence for it. (We give more evidence in chapter 5.)

The ZFEL predicts that, in the absence of selection, the number of different genes and the degree of differentiation among genes should rise spontaneously. In other words, since genes are parts of an organism or, more immediately, parts of a genome, the complexity of the genome should increase. The case is clearest for gene duplication, in which a gene is copied, by any of several mechanisms, and then—if the copy is not under selection (say, if it is functionally redundant)—mutations accumulate. Eventually the accumulation of mutations renders the copy nonfunctional, and it becomes a pseudogene (Zhang 2003). Actually, the moment of transition from gene to pseudogene is not of particular interest here, because a pseudogene is defined functionally, as a former gene that is no longer transcribed owing to the accumulation of mutations. For present purposes, what matters is only the process, the accumulation of differences in the copy—via nucleotide substitution, deletion, insertion, rearrangement, etc.—because this is what increases the copy's degree of differentiation relative to the original, regardless of whether the copy is transcribed. In other words, what is relevant to the ZFEL is the accumulation of differences. The copy becomes an ever-more-different part type in the genome, and as a result the pure complexity of the genome rises.[17] The case is a nice one in that pseudogenes produce the increase in genome complexity under near-ideal circumstances, where selection is very likely absent (or at most very weak), the zero-force condition.[18]

Now, of course, pseudogenes are also lost, sometimes by selection favoring streamlining of the genome for efficiency. But these losses are the result of separate processes, such as selection, occurring independently of the ZFEL. Thus, pseudogene loss is a reduction in complexity, but its occurrence does not contradict the ZFEL. The ZFEL claim is that the complexity of such hereditary material as exists tends to increase spontaneously. If other forces exist that tend to reduce the hereditary material, eliminating genes and reducing the genome size, complexity may decline. But the ZFEL continues to operate before, during, and after any such decline.

Notice that the focus here is different from that in most recent treatments of the duplication and differentiation notion. In molecular biology there has been great interest lately in how duplication and differentiation lead to functional novelty—called neofunctionalization—in genes (e.g., Lynch 2007a, 2007b; Lynch and Conery 2000; True and Carroll 2002; Zhang 2003; Taylor and Raes 2004), gene clusters (Garcia-Fernàndez 2005), and other coadapted gene complexes (e.g., Freeling and Thomas 2006). In one neofunctionalization mechanism, random variation in a gene copy generates mostly nonadaptive sequences—pseudogenes—but occasionally produces a new sequence that is once again transcribed to produce an RNA molecule and/or protein with adaptive properties or a new sequence that acts as a regulatory binding site for some molecule. At that point, stabilizing selection begins to act, and randomization slows or stops. Now neofunctionalization is likely a great source of novelty in evolution (Lynch 2007a, 2007b). But for the ZFEL, neofunctionalization is beside the point. When it occurs, it represents an interruption in the more general, background process of differentiation. And it is this background process that concerns the ZFEL. It may be that most pseudogenes are never neofunctionalized. Those not lost from the genome must accumulate mutations to the point where they are no longer recognizable as copies. And recognizable or not, functional or not, they contribute to the rise in genomic complexity predicted by the ZFEL.[19]

Our discussion of pseudogenes started with a gene duplication event because the comparison of a gene with its copy makes the spontaneous accumulation of differences especially vivid. But in fact, the ZFEL does not require duplications. Any two genes, even if related only distantly, will spontaneously become more different from each other over time. And the reason is that whatever inherited similarities exist between them will tend to be lost as mutations accumulate.

An objection may occur to the reader at this point. The nucleotide alphabet of a genetic message is ordinarily understood to be limited to four letters, which if true would place an upper bound on differentiation. Imagine two genes that are not under selection, and further have not been under selection for some time, so that each is maximally randomized with respect to the other. At this point, mutation accumulation would tend to destroy any chance similarities, making them more different from each other. But it would also tend to produce an equal number of chance similarities, making them more similar. Owing to the limited number of nucleotide types, many nucleotide positions will by chance converge on the same type. In that case, the differentiating tendency of the ZFEL reaches a limit, where the rates of expected gain and expected loss of similarity are equal. However, the ZFEL is not contradicted, because it predicts an increase in complexity only in the absence of other forces and constraints, the zero-force condition. Gene length is one constraint. Another is the conservativeness of the nucleotide alphabet, which is the result of other physical-chemical forces and constraints (plus selection, presumably). In the absence of such forces, it is clear that the nucleotides would vary, resulting in a fifth nucleotide type and eventually a sixth, and so on. In other words, the expectation is that randomization at the chemical level would allow the degree of differentiation between any two genes—and therefore the complexity of the genome as a whole—to increase indefinitely.

Beyond Pseudogenes. The pseudogene case makes the action of the ZFEL vivid. But our hierarchical understanding of randomness actually makes the prediction of genome randomization far more general than pseudogenes. Consider two paralogous genes, both evolving under tight selective control for different functions. Suppose that one codes for a protein that is selected for its ability to chaperone protein folding and the other for a lens crystallin selected for its ability to refract light (True and Carroll 2002). The ZFEL says that, as long as the two genes are evolving independently, they will tend to become ever more different from each other. By hypothesis, selection is present at the level of the gene but not at any higher level. In other words, we assume there is no selection acting to keep the two genes similar and none directly favoring their becoming different from each other. In our hierarchical perspective, they would be said to be evolving *randomly with respect to each other*. From the perspective of the genome as a whole, this process is indistinguishable from true randomness, and the ZFEL makes the same prediction.

We can go further yet. The two genes in question need not even be paralogous. They may be completely unrelated to each other. If each is under independent selection, the expectation is that they will change randomly with respect to each other, and therefore that they will become increasingly more different from each other. And as they do so, the complexity of the genome increases. This last case potentially brings a huge number of cases under the ZFEL umbrella. Indeed, it suggests that genomic differentiation generally, its entire history, in all lineages, could be largely the work of the ZFEL.

5 Evidence, Predictions, and Tests

The ZFEL is supported by an enormous amount of evidence, at every temporal and physical scale, at every level of organization, across biology. Two of the most compelling pieces of evidence—the increase in phenotypic diversity over the history of life and the rise in genomic complexity marked by the divergence of pseudogenes—were discussed at the ends of chapters 3 and 4. In this chapter, we give more evidence for the ZFEL, although we cannot give all of it, partly for lack of space and partly for lack of expertise. Our goal here is actually quite modest. It is to show how diverse are the lines of empirical support for the ZFEL. It is also to show how commonplace the underlying principle is in biology and therefore how well integrated it is into mainstream evolutionary thought, albeit without having been recognized or named. In all of the cases discussed, the evidence is widely known and uncontroversial, and our understanding of it is completely conventional.

Also in this chapter, we offer some novel predictions, some routes to testing the ZFEL with new experiments. The existing evidence is powerful support for the ZFEL, but some will demand more than consistency with existing evidence, the thinking being that unusually broad claims demand unusual empirical support.[1] And even for those who think consistency sufficient, it is always satisfying to see a novel prediction made and borne out by new data. Here we give the results of one small novel test we have done, showing an increase in complexity in a system under reduced selection. And we suggest other such tests.

We doubt the evidence here will be enough to convince a determined skeptic. On the other hand, we do not expect many determined skeptics. The notion that variation tends to arise and to accumulate, and that it will do so unless opposed, producing differentiation (the notion at the heart of the ZFEL), will probably strike most people raised in the Darwinian tradition as so obvious as to make empirical demonstration unnecessary. But even if so, working through some of this evidence is worthwhile, we think, to illustrate if not to demonstrate.

Diversity: Predictions and Evidence

In chapter 3, we argued that the ZFEL for diversity is the standard explanation for rising diversity in macroevolution. Here we give some evidence that it operates in microevolutionary contexts as well.

1. Nucleotide positions not under selection diversify spontaneously on account of the degeneracy of the genetic code. And duplicated genes that are functionally redundant diversify spontaneously, thus allowing for precise molecular tests of the relative effectiveness of selection versus drift (Kreitman 2000; Yang and Bielawski 2000; Bamshad and Wooding 2003).[2] Also predicted by the ZFEL is the increase in genetic diversity that accompanies the action of disparate selection pressures in distinct lineages. Every case in which selection for allele X in one population and selection for allele Y in another produces diversity at that locus is an instance of the ZFEL, provided that the populations are evolving randomly *with respect to each other* (recall the discussion in chapter 3).

2. Tissues and organs that are not under selection are more variable among individuals. In other words, they are more diverse. Notice that is a special case of the phenomenon discussed in chapter 3: tissues and organs subject to independent selection pressures—in other words, forces that are random with respect to each other—diversify. For example, selection on the bone structure of the hand in mammals produced modifications for swimming in cetaceans, flight in bats, running in horses, and grasping in primates. Here the point is the parallel one that drift too produces diversity at the tissue and organ level. And both are instances of the ZFEL.

3. Homologous characters can be maintained by stabilizing selection in multiple lineages, and yet the developmental and genetic mechanisms that underlie them may diverge over time, a process that True and Haag

(2001) call "developmental systems drift." For example, *Drosophila melanogaster* and *D. simulans* have the same pattern of thoracic bristles, shared with their common ancestor, but the underlying developmental mechanisms have apparently diverged in the two lineages. Such increases in the diversity of developmental mechanisms are manifestations of the ZFEL. Notice that this is true whether the underlying developmental mechanisms are truly drifting, varying in ways that are neutral in fitness terms, or whether they are changing under the influence of other, independent selective forces (perhaps selective co-opting of the developmental mechanism for an additional function).

4. Laboratory populations, from microbes to mice, spontaneously drift, changing randomly with respect to each other, and, therefore, diversify (Kanthaswamy and Smith 2002; Simpson et al. 1997). That is why—when genetic uniformity of laboratory populations must be maintained over space and time (for reproducibility of experimental results)—the labs responsible for preserving and supplying mouse strains cryopreserve the embryos, thus literally "freezing genetic drift in its tracks."[3] Although this diversifying process is described as "genetic drift," strictly speaking it is not. It is instead a combination of drift plus mutation—and that is the ZFEL (chapter 6).

None of these observations, models, or lines of reasoning is novel or surprising. So what is the point of the ZFEL? First, it offers unity. A heretofore-unconnected set of phenomena—ranging from the molecular predictions that govern our most fundamental methods for separating out the effects of selection from drift to the behavior of species and higher-level taxa that govern our major methods for phylogeny reconstruction—are revealed to be instances of the same underlying principle. Second, the ZFEL enables us to see a tendency in evolution that might otherwise be obscure. In the bedlam of evolution occurring at multiple levels, across myriad taxa, in disparate ecological circumstances, it is easy to lose sight of a common directional tendency—for diversity to increase—that is present always, everywhere, at every level.

Complexity: Predictions and Evidence

The pseudogene case and the more general case of gene differentiation (discussed in chapter 4) are the best evidence we know for the ZFEL in its application to complexity. But there is also evidence from phenotypic evolution.

1. Consider the study of what is called fluctuating asymmetry, or FA for short (Leamy and Klingenberg 2005). FA is concerned with the

irregular differences between left and right—for example, the small differences between right and left ribs in a vertebrate. Such differences are complexity in our sense, differences among parts. A standard assumption of FA research has been that selection favors the stability and robustness of development, the ability of development to resist the randomizing effects of environmental and genetic variation. Thus, from one generation to the next, symmetry is an indicator of that stability and robustness, and irregular differences arising between left and right reveal the power and penetrance of randomizing influences. Randomizing influences, producing asymmetrical structure, are the default expectation, the expectation when selection fails (although see Leamy and Klingenberg 2005). In effect, the ZFEL has been a fundamental assumption of the FA research program.

2. The stable asymmetries that arise in evolution are also evidence for the ZFEL. These were mentioned briefly in chapter 4 in our discussion of randomness. Here we offer them as evidence for the ZFEL. Presumably, stabilizing selection for symmetry is strong in bilaterally symmetrical organisms generally. Such organisms typically intercept and make use of the world in a symmetrical way, as much from the left as from the right. Also, left and right copies of organs may serve as backups for each other in the event of damage, and to the extent this is so, selection is expected to favor similar function and therefore similar structure. Still, many instances of directional asymmetry are known (Palmer 2005), such as that between the claws of a fiddler crab. And these are presumably the result of directional selection away from a primitively symmetrical condition. For example, the large claw in fiddler crabs seems to have been selected, independently of the smaller one, for its specialized function in sexual competition. In such cases, where asymmetry is the result of independent selection pressures for distinct functions, the two structures can be said to evolve randomly with respect to each other, and their differentiation is therefore properly an instance of the ZFEL.

3. Conventionally, differentiation in the evolution of homologous series is attributed to selection modifying subsets of the series for specialized functions. Examples include the differentiation of the limb series in arthropods, the vertebral column in vertebrates, teeth in mammals, the forewing and hindwing in insects, the forelimb and hindlimb in tetrapods, and many others. To the extent that the selective forces affecting each subset are unique, and to the extent that developmental correlations along the series were imperfect, or could be broken, each subset was modified independently of all others. And therefore, if the standard view is right, in each case the differentiation of the series as a whole—the resulting rise in its complexity—is formally attributable to the ZFEL.

A Novel Test

Here we will look at a case where the result was not known to us or to anyone ahead of time, making this in effect a novel test of the ZFEL. In particular, we will look at change on a very short timescale, comparing the morphological complexity of a group of parents with that of their newborn offspring, that is, comparing parents of laboratory animals with the variants they give birth to. The ZFEL prediction is that, absent selection and constraint, offspring will tend to be morphologically more complex than their parents. This is not a perfect test of the ZFEL, of course, because the constraints are unknown and also because selection is never totally absent. Examining newborns eliminates the effect of differential mortality and reproductive success after birth, in other words, the ecological component of selection, but there is still differential survival before birth, beginning at the moment of conception, in other words, the developmental component of selection. Still, this seems a worthwhile test, revealing as it does the effect of *reduced* selection.[4]

In the 1960s, at the Oak Ridge National Laboratory, U. H. Ehling led a series of studies on the effect of high doses of radiation on the vertebral columns of the offspring of irradiated male mice (e.g., Ehling 1965, 1966). Ehling was not interested in complexity, but his data are interpretable in terms relevant to the ZFEL. The mammalian vertebral column is a series of repeated parts, all very similar to each other but also differentiated into types. Because the parts are well bounded and cleanly identifiable, for the most part, it is an excellent structure in which to look for the effect of the ZFEL.

As discussed, in mammals, the vertebral skeleton is usually understood to consist of five vertebral types (fig. 5.1): cervical (neck), thoracic (chest, rib-bearing), lumbar (lower back, non-rib-bearing), sacral (attached to the hips), and caudal (tail). Here we treat each as a different part type, so that thoracics are one part type, lumbars another, and so on. The exceptions are the first two cervicals—atlas and axis—which differ substantially from the other neck vertebrae, as well as from each other, and so for present purposes were considered unique part types, not standard cervicals, bringing the part type count to seven.

In a series of repeated part types like a vertebral column, there are many ways for complexity to increase. Any significant variation arising in a single vertebra other than the atlas and axis adds an eighth type to the column as a whole.[5] Here are the complexity-increasing variants that Ehling (1965) reported (fig. 5.1): (1) The "dyssymphysis" of certain vertebral pieces, or elements, meaning the failure of those elements to

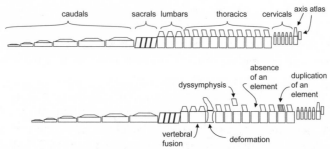

FIGURE 5.1 *Top*: A mouse vertebral column (shortened), showing the various types. *Bottom*: Examples of variants that arose in Ehling's mice.

fuse to a vertebral body, as they would in normal development. For example, in a thoracic vertebra, dyssymphysis of the neural arch transforms a normal thoracic into two new part types, the unfused neural-arch element and the vertebral body, now missing its neural-arch element. (2) The total absence of an element. For example, in one variant animal, the neural arch failed to arise in a thoracic vertebra. The archless vertebral body in this thoracic vertebra counts as a new part type in that it differs substantially from other thoracics. (3) The duplication of a vertebral element and its fusion with a vertebra. One animal had a cervical vertebra with a second neural arch fused to it. (4) The fusion of two adjacent vertebrae of the same type. The joining of the two produces a larger unit substantially different from others of the same type. In one mouse variant, two sacral vertebrae fused. (5) An asymmetry, change in size, or more generally any malformation of a vertebra. For example, one animal sported a deformation of the neural arch of one of its thoracics, generating a thoracic that was significantly different from the other, undeformed thoracics. In each of these, at least one new part type arose.

Some variations produced no change in complexity. In two variants, one thoracic was duplicated, increasing the length of the thoracic series. A part was added, but not a new part type. In another variant, there was a deformation of the axis, adding a novel part type but also eliminating a preexisting type (the old axis morphology), producing no net change in complexity. What about reductions in complexity? There were none among Ehling's mutants. Possible routes to reduction are imaginable. One would be the elimination of an entire series, such as the loss of all thoracics. Of course, if such a variant had arisen, it could well have been lethal in utero and therefore would not appear in the sample. Another route to complexity loss, one that might be observable, would be a mutation transforming a type with a single representative into a member of an existing type series, such as the conversion of the axis into a thoracic. There are other routes, but overall there do seem to be many more ways to increase the complexity of a vertebral column than to decrease it. And it is not hard to see why. A vertebral column is a massively redundant structure, like virtually all structures in biology. And the addition of random variation to a redundant structure destroys some of that redundancy, in other words, complexifies it.

Ehling divided the variants into those occurring only once, in a single animal (what he called class I variants), and those occurring in two or more animals (class II) (Ehling 1965). In class I, there were 20 morphological changes (in 10 animals)[6] in the vertebral column that could be interpreted in complexity terms, and of these, 17 were increases in complexity, 3 showed no change, and none were decreases. Results are summarized in table 5.1. In class II, the multiple-occurrence abnormalities, there were 9 cases of vertebral fusion, interpreted here as producing new vertebral types and therefore as increases in complexity. There were also 6 partial or total dyssymphyses, in which neural arches either partly or totally failed to fuse with the underlying vertebral body. Where dyssymphysis was total, the neural arch and the vertebral body elements both count as new part types. Where it was partial, the resulting vertebra was sufficiently different from others of the same type that it counts as a new type. Thus, in class II, complexity increased in every variant recorded. Complexity increases predominated.

A possible concern has to do with our classification of vertebrae into distinct types. Not all thoracic vertebrae, for example, are truly identical. And one might reasonably point out that our decisions about when a variant vertebra constituted a new type were somewhat subjective. However, the differences among vertebrae within a type are small relative to

Table 5.1. Complexity changes in 10 mouse vertebral columns

Complexity increases		
Dyssymphyses	5	
Losses of vertebral elements	5	
Duplications of vertebral elements	1	
Vertebral fusions	1	
Deformations of nonunique vertebrae	5	
Total increases		17
No change in complexity		3
Complexity decreases		0

Source: Based on descriptions in Ehling 1965 (his class I data only)

the variations Ehling observed. And therefore we think it likely that a quantitative treatment—in which complexity was measured as degree of differentiation among the vertebrae in a column, perhaps using the statistical variance or some related metric—would produce concordant results. Methods for doing this are available and have been applied in a related context (McShea 1993).

Beyond its value as a test of the ZFEL, our hope is that this case will help clarify the ZFEL claim for morphology. It shows that the seemingly radical claim that morphology should spontaneously complexify is not radical at all, that it is a straightforward extension of the pseudogene argument. As with pseudogenes, the vertebral column represents a set of well-defined parts, initially similar to each other. In both, differentiation can be assessed in an intuitively reasonable way, in the genome as nucleotide differences and in the vertebral column as differentiation among vertebral types. And in both it is easy to see how variation should, and in fact does, lead to an increase in complexity.

That said, we need to repeat that the application to a serial structure is optional. Any set of parts—not just those arranged in neat, well-demarcated series like genes and vertebral columns—is expected to become more complex. The ZFEL predicts that a set of parts consisting of, say, an ulna, a liver, and a neck will become more complex in the absence of selection and constraint. Of course, measuring degree of differentiation in sets of radically different parts is not straightforward. In many such sets, not only are the parts not well bounded but their structures are not comparable, making it hard to devise measures of differentiation. These problems are much less severe, and therefore opportunities for testing are greater, using serial homologues (McShea 1992).

Other Tests

Other tests can be imagined. One could take advantage of the presumed absence of selection on vestigial structures, structures with no known function. Consider the well-known cases of cave species with close relatives living on the surface, notably certain crayfish and fish (Fong, Kane, and Culver 1995; Culver, Kane, and Fong 1995). In many cave species, individuals spend their entire lives in the dark, which presumably reduces selection on vision and on the parts of the nervous system involved in sight. The ZFEL predicts that the eyes should become more complex, that is, the parts of the eyes should become more differentiated internally. For example, Tokarski and Hafner (1984) found variation in size and shape of rhabdomes and corneal facets among the regions of the eye of surface-living crayfish. The ZFEL predicts that in cave-adapted species, where eye ultrastructure is presumably not under selection, regional variation among these parts within the eye should be even greater.[7] The same goes for variation among other sets of parts, structural and neural, within the eye. As far as we are able to discern from the literature, no intraorganismal test for increased variation among parts has been done.[8]

The test would be imperfect, of course. One difficulty is that selection can never be completely eliminated and could bias results against the ZFEL prediction. Constraints in the form of developmental biases toward loss could do the same thing. Another difficulty is that in cave animal eyes, selection for economy in development tends to reduce the eyes, reducing the number of parts they contain along with the number of part types. A partial solution is to consider only the eye structures that remain (assessing the degree of differentiation among them), those in which reduction has not been so extreme as to overwhelm the ZFEL. A third problem is that constraints may block complexification. Even if eyes are not under stabilizing selection, they may be developmentally linked to other structures that are.

On the other hand, the developmental component of selection seems less likely to be a factor than in the case of the vertebral column in that decreases in eye complexity seem less likely to be fatal prior to birth. Also, developmental constraints can often be detected, and once detected, the structures they implicate can be avoided. Still, given all the difficulties, it is clear that no single test will be decisive. A fair test of the ZFEL would involve a number of different sets of parts in the eye. And of course a strong test would go even further, investigating part types and differentiation in a number of vestigial structures across a wide variety of taxa. A great many candidates for vestigial structures are known, providing many opportunities to test.

Complexity in the History of Life

We have reserved for last what some may think of as the most important piece of evidence, the putative trend in the morphological complexity of organisms in the history of life. One hears it said that modern organisms are more complex than ancient ones, that the history of life is a story of ascent in complexity, from bacterium to human, from "monad to man" (Ruse 1996). If so, if the conventional wisdom is right, that trend might seem to constitute powerful evidence for the ZFEL. But the issue is a difficult one. For one thing, much of the discussion of this trend in the existing literature concerns colloquial complexity and is not directly relevant to the ZFEL. For another, the ZFEL prediction for complexity at this scale is actually very different from what those familiar with that literature might think, when first considering the problem. When that prediction is properly formulated, we see that the existing evidence could be construed to support the ZFEL, but the case is ambiguous.

Trends and Trend Mechanisms. We begin with a short primer to make clear a critical distinction between two sometimes-conflated concepts: a "trend" and its underlying "mechanism." A trend is directional change in some statistic for an evolving group, usually the mean. A trend mechanism is the pattern of change underlying that trend. Consider figure 5.2A, illustrating one way that mean complexity could, in principle, have increased in the history of life. Life begins as a single species, and with the passage of time, extinction occurs (vertical lines terminate), but new species also originate (horizontal lines), and on the whole, diversity increases. The horizontal axis is complexity. For now, let us say it is complexity in the colloquial sense, ignoring for the moment that it is really unmeasurable. (We will return to pure complexity soon.)

Notice that the original species—representing the first species in existence, such as a bacterium of some sort—is quite simple, lying far to the left on the graph. Notice too that in the figure, when new species arise, they are almost always more complex than their ancestors. Most horizontal line segments are jogs to the right. In other words, there is a pervasive tendency for colloquial complexity to increase, manifest in the high percentage of right-branching lineages, what is called an evolutionary "drive" (Gould 2002). So here is how we will describe what is going on in figure 5.2A. A *trend* occurs, in that the mean increases. And the underlying *mechanism* is "driven," more precisely it is "strongly driven," in that the vast majority of changes among the lineages are increases.

Before we consider any other trend mechanisms, we need to draw attention to two other trend statistics, the maximum and the minimum. (We will need to refer to these shortly.) A trend in the maximum is the increase in the complexity of the most complex organism in existence at a given time, shown in figure 5.2B by the dotted line on the right. This might represent the putative increase in colloquial complexity from the earliest bacterium (3.5 billion years ago), conventionally considered to be simple, to modern humans (represented by the tiny vertical line segment in the uppermost right), considered by many to be the apotheosis of colloquial complexity. Another trend statistic is the minimum, the complexity at a given time of the least complex organism in existence. A trend in the minimum is shown in figure 5.2B by the dotted line on the

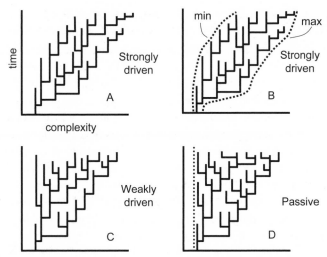

FIGURE 5.2 Alternative trend mechanisms.

left. In a strongly driven trend, the minimum is also expected to rise. The original simple species present at life's origin eventually becomes extinct, as all species do, giving rise before they go to more complex descendants and leaving a vacuum on the left side of the graph. In a strongly driven trend, simple organisms vanish, and because there are few leftward branches, they are never replaced.

Now consider the third graph, figure 5.2C. There is once again a trend, in that the mean increases. In this case, however, the drive is much weaker. More than half of the new lineages branch to the right, but many branch to the left as well. Complexity often decreases. So the trend is driven, but the rightward bias is much weaker than in A, so the mechanism is called "weakly driven." What about the maximum and the minimum? The maximum increases once again, as in a strongly driven trend, reflecting the rise from monad to man. But the minimum does not. In a weakly driven trend, decreases in complexity are common, so that when simple species go extinct, they are replaced from above, with the result that the minimum remains roughly constant. Simplicity persists.

Finally, consider a third mechanism, shown in figure 5.2D. Again there is a trend, in that the mean increases, but over most of this graph, there is no rightward bias at all. Increases and decreases happen equally often. What causes the trend is a boundary, a lower limit on complexity lying just to the left of the first species. On account of the boundary, the group as a whole diffuses to the right with the passage of time. In the actual history of life, such a boundary might represent a lower limit on colloquial complexity for living organisms, the minimum amount of complexity required for life to exist. Such a mechanism is called "passive," reflecting the diffusive nature of the spread to the right.[9]

A trend occurs in all three cases, meaning that the mean increases. But the underlying pattern of change—the trend mechanism—differs among them. In A, there is a strong tendency for complexity to increase, in C a weak tendency, and in D no tendency at all (at least away from the lower bound). For the most part, modern discussions of colloquial complexity in the history of life have taken the existence of a trend for granted, and the debate has instead revolved around the question of mechanism, of whether there has been any tendency for complexity to increase (strong or weak) or whether the trend has been the result of diversification in the presence of a boundary (passive). What is at stake here? In this debate, there has been an assumption that a driven mechanism must be connected with selection, that the discovery of a driven mechanism would imply that selection has favored colloquial complexity, on the whole, in the history of life.

A number of tests exist for distinguishing these mechanisms in real data (McShea 1996). One is based on the behavior of the minimum. The minimum does not change in the weakly driven and passive mechanisms, but an increasing minimum identifies a trend as strongly driven. (In contrast, the behavior of the maximum is not very informative. The maximum increases in all three cases.) The question of mechanisms was raised, implicitly, by Stanley (1973) in a discussion of trends in body size, developed further by Gould (1988), placed in the context of complexity by McShea (1993, 1996), and popularized by Gould (1996) in his book *Full House*.[10]

Colloquial Complexity: The Evidence. Has there been a trend in the mean for colloquial complexity? It will surprise some to learn that all of the evidence for a trend is impressionistic. The reason is that, as we have said, colloquial complexity has never been operationalized. It has never been measured in any objective, repeatable way, not in a single species. The entire debate about a trend, including Gould's treatment in *Full House* of the rise of complexity—or what he called "excellence"—is predicated on the assumption that our impressions about organismal complexity reflect some deep but ineffable underlying reality. Bacteria *seem* simple to us, protists *seem* more colloquially complex, the earliest animals *seem* even more colloquially complex, and humans *seem* the most colloquially complex of all. Beyond these impressions, however powerful and compelling they may be, we really have no evidence.

But let us take for granted that a trend has occurred. Perhaps our impressions are based on direct perception of some kind, a gestalt, that reflects a real increase in something—call it colloquial complexity—over the history of life. If so, what can we say about the mechanism, about the existence of an upward drive? Well, the first thing to say is that the rise from bacterium to human is a rise in the maximum and therefore tells us nothing about the underlying trend mechanism. The maximum rises in all three cases in figure 5.2. In contrast, the behavior of the minimum does tell us something. There is a widely shared understanding that the first organisms must have been simple and that simplicity has persisted—in particular, that modern bacteria are just as simple as ancient ones.[11] If this is right, then we can rule out a strongly driven trend.[12] And thus, if there has been any drive toward increasing colloquial complexity in the history of life, it has been a weak one. And the possibility exists that—if the trend has been passive—there has been none at all.[13] About colloquial complexity, this is pretty much all that can be said.

The Special Formulation of the ZFEL Predicts a Strongly Driven Trend! We are about to return to the ZFEL, but first we need to make clear that we are changing the subject, from colloquial complexity back to pure complexity. The ZFEL has nothing directly to do with colloquial complexity. It has nothing to do with intelligence, sophistication, functionality, or "excellence." Nor is it concerned with complexity in the sense of hierarchy, with the rise in number of levels of nestedness, with levels of selection, or with the so-called major transitions in evolution (Maynard Smith and Szathmáry 1995; Marcot and McShea 2007). It has nothing to do with evolutionary progress in any sense (Rosenberg and McShea 2007). Any or all of these may be correlated with pure complexity, perhaps even deeply and causally connected with it, but no such correlations have been demonstrated, and in any case they are not our concern at this moment. Thus, if some of our conclusions in this section seem a bit odd, or at odds with conventional thinking, it is because we are talking only about pure complexity: part types and differentiation.

Among the trend mechanisms shown in figure 5.2, which does the ZFEL predict for pure complexity in the history of life? Consider this: the individuals of every species consist of some set of parts. The ZFEL predicts that in the absence of selection and constraints—that is, in the special formulation—there is, in every species, a tendency for those parts to become more differentiated over time. Reversals can occur. Parts can by chance become more similar to each other. But mostly they will not. In other words, the special formulation of the ZFEL predicts an increase in complexity in almost every species, in almost every lineage, and therefore a strongly driven trend, like that in figure 5.2A.[14] Interestingly, notice that the ZFEL challenges the standard assumption that a strong drive would be evidence for selection. Strong selection predicts a strong drive but so does the ZFEL, and the ZFEL does so in the total absence of selection favoring complexity.

It might seem, especially to those familiar with the literature on trends, that we have made a mistake here. The ZFEL points to a diffusion of genes in a sequence space or the diffusion of part morphologies in a phenotype space. And a passive mechanism also relies on diffusion (fig. 5.2D). So it might seem that the ZFEL prediction should be a passive mechanism, not a driven one. But this argument is based on a confusion of conceptual spaces. Figure 5.2D shows a diffusion of *species* in a *complexity space,* while the ZFEL predicts diffusion of *parts* in a *phenotype space.* Suppose we represent the morphology of some species in a space with axes representing the dimensions of a series of parts. The ZFEL says that, in the absence of selection and constraints, those parts are expected to move away from each other, on average, with the result that the variance among them—their complexity—increases. If the ZFEL is right, this will be true of the parts in every species, which means that in a complexity space, where each species is represented by a point corresponding to the variance among its parts, almost all species are expected to become more complex.[15]

Consider the blowing-leaves analogy again. The ZFEL says that the leaves in a windy yard will diffuse. But this is true of the leaves in every yard, which means that, over a number of yards, the dispersion of leaves should increase in almost all of them. The leaves do diffuse in actual space, but their spatial variance—plotted in a complexity space—is driven, not diffusive.[16]

Pure Complexity: The Evidence. Surely, one might think, there has been a trend in pure complexity. Surely a modern organism has on average more part types or greater differentiation among parts than an ancient one. In fact, we have some data on this. Pure complexity is measurable, unlike colloquial. In a study of number of part types and gene numbers in modern bacteria and modern protists, Marcus (2005) found that the protists have more part types, on average. If we assume that the complexities of the modern groups are fair proxies for those of the ancients, then mean complexity undoubtedly increased in the transition from bacterium to protist, some 2.0 billion years ago. Beyond this, however, we have no evidence for a trend. No one has attempted to measure the average pure complexity of the full range of modern organisms.

And indeed there is a major barrier to measuring the complexity of a bacterium in a way that makes it comparable, and averageable, with a multicellular metazoan, say, an insect. It is the problem of identifying a shared level of organization. We cannot count part types or measure differentiation at the tissue or organ level, for example, because single bacterial cells exist entirely at the cellular level. They have no tissues or organs in the same sense—at the same level of organization—as an insect. To make a meaningful comparison, to make averaging possible, one needs to count part types or measure differentiation at a level at which both have parts.

Suppose we turn to the molecular level, the thinking being that molecules are parts shared by all organisms. But the number of types of molecules is essentially unknown in both bacteria and insects. We can count gene types, but genes types are a small subset of the total population of molecular species. We might assume that number of genes is well correlated with number of gene products, but that has not been demonstrated, so far as we know. And in any case, it is at least questionable whether the number of molecular types in an organism is well correlated with number of gene products. Many molecular types must come from an organism's environment. Finally, it might seem reasonable to suppose that an insect has more molecular types simply because it is so much larger, but we have to acknowledge the in-principle possibility that the insect molecular complement might contain vast redundancy, perhaps enough to offset its larger size. Even at the molecular level, essentially nothing is known about a trend in pure complexity.

Our concern here is with complexity at the scale of life as a whole. But there is some evidence from the metazoans that is worth noting. Valentine, Collins, and Meyer (1994) have documented a trend in the maximum number of cell types over the history of animals. Their data show a rise from the earliest sponges, with just a few cell types, to humans, with over 200 types. Further, we can say that the minimum probably has not changed. The original animal ancestor, sometime before the Cambrian, more than 540 million years ago, was probably a protist, with a single cell type. And there is a modern group, the living sister group of the animals, the choanoflagellates, some species of which exist vegetatively in groups of cells that are all of the same type. Thus, the modern minimum and the ancient minimum—a single cell type—are the same, which is at least consistent with a constant minimum.[17] If the minimum has in fact been constant at one, the combination of a rising maximum and a constant minimum suggests a rising mean, since means are necessarily sandwiched between minimum and maximum. Of course, recall that we are at the wrong scale here. A trend in all life does not follow from a trend in metazoans. Decreasing trends in other multicellular groups, or even in protistan groups, could negate or overwhelm the trend in animals. Still, this evidence—combined with that from Marcus's study—is suggestive of a rising overall mean.

Let us accept for the sake of argument that there has been a trend in the mean for all life. Does such a trend provide support for the ZFEL? By itself, without knowing anything about the trend mechanism, it provides very little. Trends in means can result in a number of different ways, only one of which is predicted by the ZFEL. That was the point of figure 5.2. So what about the trend mechanism? Is there any evidence for the strong drive predicted by the ZFEL? What evidence we have points to the opposite. Again, no one has measured the pure complexity of ancient bacteria and compared them to the moderns, but modern bacteria are at least superficially similar to the ancient ones, and if we assume that both are simple in the pure sense (in addition to being simple in the colloquial sense, as we assumed earlier), then the complexity minimum has remained stable over the history of life. And a stable minimum is the expectation for a passive or weakly driven mechanism, not the strong drive predicted by the ZFEL.[18] In other words, a stable minimum is not the expectation of the special formulation of the ZFEL. It is consistent with the general formulation, of course, because there are many ways in which the zero-force assumption might be violated. In other words, it could be that something is opposing the strong drive the ZFEL predicts.

We conclude that what is currently known about the history of life offers little evidence for the ZFEL for complexity. A long-term increase in the mean has not been demonstrated, but if it in fact occurred, it would be consistent with a number of possible underlying mechanisms. The ZFEL predicts a strong drive, but no such drive has been shown, and indeed the stable minimum suggested by impressionistic assessments argues for the opposite, a weak drive or none at all. And this finding in turn suggests that, if the ZFEL is true and if it has acted in the history of life, it has been opposed by constraints or countered by some force.

Some Speculation. Let us take for granted that the ZFEL is true, that a trend in the mean occurred over the history of life, that the minimum has been stable at a low level, that the strong drive predicted by the ZFEL is not manifest, and that therefore some constraint or force has acted in opposition. Consider an analogy. We have a physical theory, a theory of gravitation that predicts that the helicopter we see hovering in the distance should fall to the ground. It does not, so we infer that some opposing upward force must be acting, perhaps a downward wind generated by the helicopter. Similarly in evolution, theory identifies a strong complexity-increasing tendency, the ZFEL, acting everywhere and always. But we do not think we see any such strong tendency in the history of life. We infer that something must be opposing it.

We can imagine two sources of opposition. First, it might be that the drive predicted by the ZFEL is overpowered by some constraint. One candidate is a bias built into the structure of development, discussed in chapter 4.

A second possibility is that change in complexity is controlled entirely by selection, overwhelming the ZFEL, and that selection has favored increases and decreases in complexity about equally frequently (McShea 1993). Or, along the same lines, it could be that the ZFEL-driven increase is overpowered by the on-average disadvantages of complexity; in other words, there is a rough balance between the ZFEL and selection against complexity. These may sound implausible, initially, if only because the on-average *advantages* of complexity have been a persistent subtheme in Darwinian theory. Darwin himself suggests that complexity increase may be favored, in general, on account of the advantages of the division of labor among parts that it affords (Darwin [1859] 1964, 1987). And in recent decades, a variety of other selective rationales have been offered. Saunders and Ho (1976) argued that complexity is favored, on average, because part losses are more likely to disrupt function than part gains.[19] Bonner (1988) has argued that selection favors large body size, on average, and larger organisms typically require greater complexity (see also Bell and Mooers 1997; Bonner 2004). For example, small organisms have larger surface-area-to-volume ratios and therefore can use diffusion to supply their tissues with oxygen. But large organisms have lower ratios and so require circulatory systems, which in turn require specialized part types, leading to more complexity. And McCarthy and Enquist (2005) argue that greater complexity is demanded by increases in a combination of body size and what they call metabolic intensity because new parts perform new functions and more energy is required to perform those functions as well as to integrate them with preexisting ones.

On the other hand, simplicity is expected to be favored often too. Conventional thinking is that parasites generally lose part types in the transition from a free-living existence (although see Brooks and McLennan 1993), the usual rationale being that fewer part types is an adaptation for efficiency in the resource-rich environment of the host, where less mechanical and metabolic machinery is required. Also at least some of the so-called meiofauna, small animals living in the interstices between particles (such as sand grains), may be secondarily simplified from larger, more complex ancestors. And reduction in complexity is likely directly connected with the reduction in size (Bonner's argument, inverted), at least in those taxa that are derived from larger, more complex ancestors. It has also been suggested that complex species might have greater extinction probabilities, if complex means ecologically specialized and greater specialization means reduced ability to adapt to environmental change (Schopf et al. 1975). And complex designs may often be unstable in an engineering sense, making them on average less fit (Wimsatt and Schank 1988). There is also the notion, originating with R. A. Fisher, that more complex organisms are likely to be more unstable in an evolutionary sense. In other words, a small variation in a complex organism is more likely to be deleterious than a small variation in a simple one (Orr 2000).

This analysis just scratches the surface. We have not even touched on the rich and potentially relevant literature on modularity, pleiotropy, evolvability, and so on, nor on the recent work using "digital organisms" on the effect of selection on complexity (variously defined). But it suffices to make the point that, on the question of the expected effect of selection on complexity, theory does not speak with one voice. And therefore, the notion that some constraint or force is overpowering the ZFEL is a live possibility.

Final Thoughts on the Putative Complexity Trend. The ZFEL for diversity seems well supported by data from the smallest to the largest hierarchical scales and from the shortest to the longest timescales. However, the ZFEL for complexity is more problematic in that the prediction it makes for the whole history of life on Earth seems not to be well supported by available data. This raises two apparent problems. First, it might seem to contradict our earlier point that complexity and diversity are the same thing viewed from adjacent hierarchical levels. How could diversity increase over the history of life but not complexity? But this problem is only apparent. In the foregoing discussion, we focused on *organismic* complexity, which for complexity is the level that most biologists are interested in. But the appropriate level to compare with the diversity history of life is the clade, not the organism. It is the "complexity" of higher taxa, meaning the number of and disparity among lower-level taxa within them (the quotation marks reflecting the fact that biologists do not ordinarily use the word in this way). And, consistent with the finding of the last chapter, the "complexity" of higher taxa surely does increase over the history of life. Second, the apparent absence of a strongly driven ZFEL-like increase in organismic complexity might seem to contradict the central prediction of the ZFEL; indeed, it might seem to contradict our whole outlook. But the point of this book is not that the conditions for the special formulation of the ZFEL are met at all levels and at all times. They certainly are not. On the other hand, we argue, the general formulation is always and everywhere applicable. It gives us the background against which hypotheses about forces and constraints can emerge. In this case, it raised a new possibility: some force or constraint may be thwarting the ZFEL tendency. Selection may be working against complexity.

6 Philosophical Foundations

In the previous chapters we have tried to convince the reader that the ZFEL is true. Here we step back and pursue more philosophical topics: What makes the ZFEL true? What justifies the scientific/philosophical stance we take in which the ZFEL becomes fundamental?

Answering these questions will involve exploring the connections of our ideas to those in the theory of genetic drift in evolutionary biology and to basic probability theory. The connections to probability theory prove to be both more surprising and deeper. So deep that we will end up arguing, contrary to every philosophical tradition of which we are aware, that probability theory is the reductive foundation of evolutionary theory (so much for physicochemical reductionism). But we have a ways to go before we get to such a grand conclusion. First we must deal with drift and its relation to the ZFEL.

The Principle of Drift and the ZFEL (Diversity)

Ernst Mayr (1963) argued that one of the greatest philosophical breakthroughs of the Darwinian revolution was population thinking, that is, taking populations as real, not as mere artifacts of some mathematical operation performed on their component individuals. This breakthrough would not have been possible without simultaneous conceptual and mathematical developments in statistics. Indeed, many of the pioneers of late-nineteenth- and twentieth-century statistics were biologists, driven by biological problems. In trying to make sense of heredity, Francis Galton and his protégé Karl Pearson gave us many of the standard tools of statistics, for example, regression to the mean, the Pearson distribution, and the chi-square test. It was only with the development of these tools that Darwin's theory was susceptible to experimental test (Weldon 1898). In the early twentieth century, R. A. Fisher developed the analysis of variance and maximum-likelihood tests (just to mention two of his best-known contributions to statistics) in his efforts to understand evolutionary genetics.

Contemporary evolutionary science is, of course, thoroughly statistical. One area where this is particularly obvious is in the study of genetic drift. "Drift" is typically defined in terms of the statistical concept of sampling error. Shortly, we will offer a precise definition of drift, but for now let us illustrate the idea using a common example. Consider a large urn filled with balls of different colors—for simplicity, let us say red and black. These colors occur with a frequency of p and q respectively. Now, assuming the balls are thoroughly mixed, we sample one ball, record its color, replace it, remix, and sample again. The probability that we will get a red ball is $\Pr(\text{red}) = p$; while $\Pr(\text{black}) = q$. That is, the probability that we will get a particular color on a particular draw just equals its relative frequency in the urn in this particular experimental setup. Sampling with replacement keeps the probabilities constant and so is a nice simplifying assumption. However, when we try to use the urn as an analogy for biological sampling (e.g., gametic sampling in reproduction or parental sampling when a population goes through a bottleneck), sampling with replacement is often not an accurate analogy. It would be better to think of sampling without replacement. In this case, the probabilities have to constantly be updated in terms of what has already been sampled. But, so long as the sample size is small relative to the size of the population being sampled, the case of sampling with replacement provides a good approximation of the actual probabilities—and it will suffice for our purposes.

Now suppose the urn contains 10,000 balls and we sample only 4. What is the likelihood that the frequency distribution in our sample will match the frequency distribution in the urn? That question cannot be answered without specifying the values of p and q. As can easily be seen by using the probability calculus, the closer the values of p and q are to each other, the more likely it is that we will get a result close to the true population frequency and vice versa (except for the extreme case where one of the two frequencies is 0; in which case our sample will exactly match the true frequency).[1] Consider the case of $p = q = 0.5$. We sample four balls. The possible outcomes are as follows, with their associated probabilities:

Pr(all red) = 0.0625
Pr(3 red, 1 black) = 0.25
Pr(2 red, 2 black) = 0.375
Pr(1 red, 3 black) = 0.25
Pr(all black) = 0.0625

The most probable outcome, the one that will occur most often in a long sequence of such trials, that is, the *mode*, is 2 red and 2 black. That outcome is also the *expected*, or *mean*, outcome, that is, the one that corresponds to the overall frequency of the two types in the urn. (The mode and the expected outcome will not generally be the same. They are the same here because the values of p and q have been set to 0.5.) Notice that, while the expected outcome is indeed the modal outcome, it occurs in slightly fewer than 4 out of 10 trials, or conversely, outcomes that deviate from the expectation occur in slightly more than 6 out of 10 trials. If we think of drift as *any outcome that deviates from the expected outcome*, then we can see that in this setup (i.e., very small samples from a very large original population) drift is highly likely (Brandon 2005). Thought of this way, drift is a highly abstract idea, not to be identified exclusively with genetic drift.

Now recall the simple model introduced in figure 2.1 (reproduced here as fig. 6.1). A particle starts at point 0 in space and moves to the right with Pr = 0.5 and to the left with Pr = 0.5. And it behaves according to this same rule in each unit of time. If we follow a single particle through space through four steps and ask where it is at time t_4, we end up doing the exact same bit of probability calculus we did above, getting, naturally, the same numbers. Just as there were five possibilities above in our sample of four balls, here there are five possible positions for our particle on the x-axis. It could be at $x = 4$, $x = 2$, $x = 0$, $x = -2$, or $x = -4$. The probabilities associated with each of these possibilities are the same as above for the expected result, $x = 0$ having Pr = 0.375. Thus, a single particle can be thought of as drifting through space. Alternatively, a single particle can be thought of as tracking the allele frequencies (p and q) of two neutral alleles that start at a 50/50 ratio as a population samples from these two alleles each generation, represented by each time step.

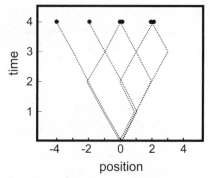

FIGURE 6.1 The increase in variance over time in an ensemble of six particles.

In other words—and here is the point—change in any of these contexts counts as drift because ultimately drift is simply a consequence of probabilistic sampling, and nothing more (Brandon 2005).

Are there any lawful regularities regarding drift? There are, and the theory that has been developed has articulated a number of them. For instance, to take the example above, if one were to start with an ensemble of populations, each with two neutral alleles A_1 and A_2 at frequencies p and $q = 0.5$ respectively, then approximately 50% of those populations will become fixed for A_1 while the other 50% will become fixed for A_2. Accurate predictions can even be made regarding the time to fixation based on population size (Roughgarden 1979). A little reflection on these predictions will reveal that they are entirely based on probabilistic reasoning.

Aside from these well-known regularities involving drift, is there something more general that can be said about it? We think so. Brandon (2006) stated what he termed the principle of drift:

> (A) A population at equilibrium will tend to drift from that equilibrium unless acted on by an evolutionary force.
>
> (B) A population on evolutionary trajectory t will tend to depart from that trajectory (in either direction or magnitude or both) unless acted on by an evolutionary force.

Like the ZFEL, we do not think of the principle of drift (PD) as a novel biological discovery; rather, we think of it as a useful generalization and systematization of much that is already known. Once stated, its truth is rather obvious. Populations left alone drift. The null microevolutionary expectation is change. For instance, what should we expect to happen if two laboratories receive shipments of genetically identical strains of mice from a reputable supplier and then separately maintain these strains for 100 generations? Clause A tells us that we should expect the two laboratories' strains to differ from each other and both to differ from the original strain. Clause B tells us that there is nothing like Newtonian inertia in evolution. For instance, imagine two originally identical laboratory populations subjected to identical directional artificial selection for some trait variant and then, after some number of generations, we stopped the artificial selection. Let us imagine that the two populations had followed identical (or nearly so) trajectories (through morphospace or genotype space; it does not matter) throughout the experiment. Now what happens? The two populations continue to move though morphospace or genotype space, but not on their same trajectories, as can be seen by the fact that they soon differentiate from each other. Again, this is obvious to any evolutionary biologist.

The PD Underlies the ZFEL for Diversity. We now turn to the main issue of this section and give it a preliminary answer. What is the relationship of the principle of drift to the ZFEL? First consider diversity. The PD can be thought of as governing the default tendencies of the means of populations. That is, each particle in our simple model in figure 6.1 can be understood as a population, movement of a particle is a change in the mean for that population, and the PD governs the expectation for the behavior of that mean. The PD governs what each particle, each population, does when not affected by some net evolutionary force. It drifts.

Now the ZFEL has to do with variances, not means. It says that variances tend to increase with time. Diversity is an aspect of variance. So if we have an ensemble of populations, each governed by the PD and therefore each with a drifting mean, the variance of the distribution of the means is expected to increase probabilistically through time.[2] Relating this back to our simple model in figure 6.1, each of the single points in that figure can be understood as the drifting mean of a population. Or to put it another way, the ZFEL governs the second moment of the distribution of means, each acting according to the PD. Multiple independently drifting means produce an increasingly diffusing variance, that is, increasing diversity.

The PD Underlies the ZFEL for Complexity. Parts drift in much the same way that population means do, although the language in which we describe it will be different. For any given part in a parent individual—say a cell, a tissue, or an organ—the same part in its offspring will tend to be different. Conventionally in evolutionary biology, attention would be focused on the genetic component of variation, and the differences would be interpreted as the result of imperfect sampling of genes. The imperfect sampling would be understood to be the result of mutation and—in sexual species—recombination. But here we will frame it more generally, remaining agnostic about the underlying mechanisms of inheritance and development. Instead, we will say that the difference is the result of imperfect sampling of the heritable causal factors that underlie the generation of the part.

So in figure 6.1, each point represents a part, and the location of that point along the horizontal axis at any time represents its phenotype. Further, we can understand that part's phenotype as the mean of a distribution of possible phenotypes that could be produced by the underlying heritable causal factors. Typically, that mean will correspond to the parental phenotype, hence the expectation that offspring parts will look like parent parts. In any case, each offspring draws from the distribution of possible phenotypes. And as a result of imperfect sampling, the phenotype of the part will tend to drift. So far, this is really no more than a formal way of saying what Darwin labored to show and what every biologist now knows. From parent to offspring, parts vary. Left alone, they drift. What the ZFEL adds is that—to the extent that parts drift independently of each other—the variance among them will tend to increase. Multiple independently drifting parts result in ever-increasing variance among them. And increasing variance among parts is increasing complexity.

Alternative Routes. Importantly, drifting means is not the only way to produce the pattern in our simple model. There is also the alternative discussed in chapter 2. Suppose that each particle represents a population, and each population is moving under the control of selection, but the selective forces on the populations are independent of each other at any given time and also change independently in time. If the horizontal axis were body size, then perhaps one population is selected to get bigger, a second population is selected to get smaller, a third to get smaller, and a fourth to get bigger. Then, in the next time step, suppose the first is selected to get smaller, the second smaller also, the third bigger, and so on, so that movements of populations along the size axis are uncorrelated. In this scenario, the movement of each population is governed by selection and is therefore not random, but collectively the movements of the populations are *random with respect to each other.* The same argument could be made for complexity. The particles could be parts within a single organism, each subject to selection for some unique function. In that case, the changes that occur in each part from generation to generation are not random, but to the extent they are uncorrelated they are random with respect to each other. In either case, parts or populations, the expectation is an increase in variance. And the general point is that the increase in variance of the ensemble can be produced either by true randomness in the behavior of the components or by thoroughly deterministic randomness-with-respect-to-each-other. The effect is the same.

It is interesting to note that once the randomness-with-respect-to-each-other is introduced at some biological level, the next level up "perceives" it as true randomness. The result is diffusion, and whatever higher-level effects that diffusion might have will be unaffected by any lower-level metaphysical distinction. The higher-level process does not care. In this way the ZFEL is an autonomous statistical law. It seems to us that the sense of autonomy used here is stronger than (and more interesting than) that introduced by Hacking (1990). That is, its stochastic character is completely independent of the deterministic or indeterministic nature of the dynamics of the individual "points" in the ensemble that the ZFEL governs.

A consequence is that, while the ZFEL can be understood to be the result of underlying drift, it need not be. In other words, the ZFEL does not *reduce* to the PD in any of the philosophical senses of that notion. The ZFEL cares only about the phenomenological pattern of diffusion, not about how it was produced. So let us say that the PD underlies the ZFEL in many cases but not necessarily all. Put another way, if one thinks of the PD as describing a causal mechanism (see the section below on drift as a casual mechanism), then the ZFEL does not reduce to the PD. If, on the other hand, one thinks of the PD as merely describing a phenomenological pattern, then the ZFEL does reduce to it.

Sewall Wright may have been the first to appreciate this consequence of drift. Multiple subpopulations, each drifting according to its own dynamic, provided the random variation on which group selection could work in Wright's shifting-balance model (Wright 1948). Drift is not the only mechanism for introducing random variation. At the base molecular level a number of mechanisms exist, none of which would be classified as drift—for example, point mutations, frame shifts, insertions, deletions. It seems likely that some of these mechanisms are truly random, as they are governed by quantum events (Ehrlich and Wang 1981;

McFadden 2001; Glymour 2001; Stamos 2001). But, again, all that is required by the ZFEL is that they be random-with-respect-to-each-other. Each of these molecular mechanisms produces mutations that are random in this sense. With non-DNA-based inheritance (e.g., cultural transmission), there are again mutation-like mechanisms for introducing random variations (think of the telephone game). Since we are interested in universal biology, we do not wish to tie ourselves too closely to known Earthly mechanisms. But the main point in this section is that drift is one very general way of producing random variation.

The Levels of Drift

Biologists tend to think of drift as genetic drift. Individual alleles, either neutral or near neutral, drift in frequency with respect to each other. That was the theory that Kimura introduced in 1968 (also see Kimura 1983). But in contemporary molecular biology we need a more expansive view of drift. Is there only one molecular unit that drifts? No. The third position in codons tends to be degenerate in the genetic code so that substitutions there are often without effect in coding regions. This is in contrast to the first two positions. This difference has given rise to one of the most powerful molecular tests for selection versus drift hypotheses. If, say, within a given population, one finds much more variation in third-position sites than in the first- and second-position sites in a given genetic region, one concludes that selection is acting there (directing change in a particular way or constraining change). But if there is no difference between third position and the first two, one concludes that selection is not constraining that genetic region (and so the null hypothesis is more likely—i.e., drift) (Kreitman 1991, 2000; Kreitman and Hudson 1991; McDonald and Kreitman 1991). So drift occurs at a fairly constant rate at third-position sites in DNA.

Such sites are nested within "genes," and we know that some of them are drifting. Given that the rate of neutral mutations in such genes need not be, and is unlikely to be, the same as the rate of third-position, "silent" mutations, genes will drift according to a different dynamic. Thus, there are at least two levels at which drift is generally acknowledged to occur, and they are at least somewhat independent of each other. This should not disconcert anyone. It is now generally acknowledged that selection occurs at more than one level, and that selection theory was greatly advanced by its generalization to multiple hierarchical levels (Brandon 1982, 1990; Lewontin 1970). This book is thoroughly hierarchical in its approach. So it should not surprise the reader to see us present a general hierarchical theory of drift.

Necessary Conditions for Drift at Any Level. Specifying the necessary conditions for drift turns out to be straightforward. We can simply follow the literature on levels (units) of selection, at least to start with. Lewontin (1970) argued that anything that satisfied the following three conditions (Darwin's three conditions) was a unit of selection:

> 1. Variation: There is variation among entities within a reproducing population.
> 2. Heredity: This variation is (to some degree) heritable; that is, offspring resemble their parents more than they do the population mean.
> 3. Differential reproduction: Some variants produce more offspring than others.

First, it is important to note that this is not a recipe for evolution by natural selection. Natural selection occurs only when there is a causal connection between the trait variant and reproductive success. Clause 3 above says nothing about such a connection. Of course, a regular and repeatable correlation between a certain trait variant and reproductive success would lead one to seek some causal connection (direct or indirect), but clause 3 says nothing about regularity or repeatability. Clauses 1–3 are perfectly compatible with drift.

Thus, we could use Darwin's three conditions to give us necessary, but not sufficient, conditions for some set of entities being subject to evolution by drift. (These conditions are insufficient, of course, because they can be satisfied and yet reproduction can proceed according to probabilistic expectations, in which case no drift will occur.) Darwin's three conditions, as stated here, work just as well for stating the bare bones of a general theory of drift as they do for a general theory of selection.[3]

Higher-Level Drift. Here we will argue that Darwin's three conditions give us necessary, but not sufficient, conditions for drift to occur at a given level of biological organization, not just the molecular level but at superorganismic levels as well. To get there, we need to start with selection. Selection at levels of organization above the organismic has been controversial, but much of the controversy has been based on conceptual confusion, not on a disagreement about the facts. A clearly articulated hierarchical theory of selection promises to place the controversy where it should be in empirical science—on issues that are subject to observation and experiment. We are getting there, even if progress is frustratingly slow. Fortunately, one of the major issues in dispute in the debate over levels of selection is irrelevant to the levels of drift, thus making a hierarchical theory of drift considerably simpler. That issue concerns the idea of ecological interaction. For something to be a level of selection, the entities at that level must be ecological *interactors*—at least, so goes the dominant strain of selection theory (Brandon 1990; Hull 1981; Lloyd 2001; Sober and Wilson 1994). As pointed out above, mere differential reproduction does not imply selection. It may occur by chance alone. Or it may occur at a given level because that level is nested within a higher-level entity that itself is subject to selection. For example, Brandon (1982) has argued that in standard cases of organismic selection one can show that selection occurs at the organismic level and not at the genic by using the probabilistic notion of *screening off* (Reichenbach 1949; Salmon 1971, 1984). By definition, A screens off B from E if and only if $\Pr(E, A$ and $B) = \Pr(E, A) \neq \Pr(E, B)$. That is, the probability of event E given both A and B equals that of E given A but does not equal that of E given just B. Consider a standard case of organismic-level selection such as selection for cryptic coloration by visual predators. Let E be a variable standing for a certain level of reproductive success, let A_1 stand for the cryptic phenotype, and let B_1 stand for the genotype that under normal conditions produces the cryptic phenotype. Clearly, for a given value of E, A_1 screens off B_1. (If one doubts this, one can see it by doing phenotypic manipulations.)

Organisms are clearly interactors. For group selection to work, groups have to be interactors. This does not seem implausible. But what about even higher-level entities, species or clades? Some have thought it highly unlikely that such entities could ever be ecological interactors. They are simply the wrong sort of thing—they are genealogical entities not ecological entities. They are spread too thinly across space and time and thus do not experience the sort of consistency of environmental pressures that selection requires (Damuth 1985). We take no stance on this issue here. Our only point is that it is irrelevant to drift. Drift does not require ecological interactors.

The hierarchical theory of drift has no need to pick out anything like an interactor. Consider population bottlenecks. In such cases the parents for future generations are sampled, let us suppose randomly, from an initially large population. They are sampled as whole organisms. But when they are sampled, so too are the parts within them—for example, their organ systems, organs, and tissues if they are multicellular. So too are the chromosomes that are contained in them (a nonrepresentative subset of the whole population of chromosomes). So too are genotypes. So too are genes. And so on down to individual nucleotides. Drift happens at all of those levels, and none is particularly privileged in this scenario.

Now back to species and clades, the stuff of macroevolution. Whatever doubts one may have about selection acting pervasively at these levels, there is no doubt that there is *sorting* at these levels, that is, differential survival and/or reproduction (Jablonski 2008). Thus, there is plenty of room for the PD to work at these levels as well.

Newton, Hardy-Weinberg, and Zero-Force Laws

We call the ZFEL a zero-force law in part to make a connection with Newton's first law, the principle of inertia, our best exemplar of a zero-force law. Here we make that connection explicit, by giving what one might call a "Newtonian" formulation of the two clauses of the principle of drift:

> (A) A population at rest will tend to start moving unless acted on by external force.
> (B) A population in motion will tend to stay in motion, but change its trajectory, unless acted on by an external force.

The formulations are zero-force in the sense that they tell us what the population will tend to do when no external force acts. But, of course, although the language in these formulations is that of objects in motion, where the objects are now biological populations, the claims of the PD are decidedly non-Newtonian. The trajectory of populations changes unless acted upon by a force.

The comparison with Newton's first law is not only apt, we think, but useful. Knowing what happens when no net force acts on an object is a necessary precondition for applying the rest of Newton's laws. To calculate the resultant of an applied force, one needs to know what the object is expected to do in the absence of that (or any) force. And for a population, to calculate the resultant of an applied force, such as selection, one needs to know what the population is expected to do in its absence. An equivalent way of thinking about the zero-force law, and a way that may be more in keeping with most biological practice, is to think of it as giving the appropriate null hypothesis. That is equivalent because a null hypothesis just tells you what would happen if nothing special were going on.

The Hardy-Weinberg Law. The Newtonian analogy is not new to evolutionary biology. Biologists have long thought of the Hardy-Weinberg law as the analog of Newton's first law in evolutionary biology (or, at least, in evolutionary genetics; see, e.g., Ruse 1973; Sober 1984; Lynch and Walsh 1998). Quite recently, there has been a heated debate among philosophers of biology about the validity and usefulness of this Newtonian analogy (Brandon 2006; Matthen and Ariew 2002; Stephens 2004; Walsh, Lewens, and Ariew 2002). Although we end up endorsing the analogy, we will show that the analogy is more properly with the PD and the ZFEL and not with Hardy-Weinberg. We do not think that the Hardy-Weinberg law is a zero-force law. We do not think that it provides appropriate null hypotheses.

The standard Newtonian paradigm is to think of stasis as the null expectation. And that is what we teach when we teach the Hardy-Weinberg law, which basically states that in the absence of evolutionary forces a population reaches both a genic and a genotypic equilibrium in a single generation and stays there until some force perturbs it.

There is an important philosophical sense in which the Hardy-Weinberg law is no law at all. Beatty (1981) has made this point. He has argued that this so-called law depends on derived evolutionary conditions, diploidy and sexual reproduction, and so is not even true throughout the history of life on this planet, much less a part of universal biology. That is to say, it is much more of an accidental generalization than a law. This is in contrast to the ZFEL, which we do take to be a part of universal biology. But we are not going to dwell on that point here. Rather, our main point here is that the Hardy-Weinberg "law" gives exactly the wrong null expectation. If Beatty's point was made with a philosophical scalpel, ours is made with a ten-pound sledgehammer.

There are two importantly different statements of the Hardy-Weinberg law:

> H-W$_1$: If a population exists with two alleles, A_1 and A_2, with frequencies p and q respectively, then in a single generation the population will settle into genic and genotypic equilibrium with gene frequencies p and q and genotypic frequencies $A_1A_1 = p^2$, $A_1A_2 = 2pq$, and $A_2A_2 = q^2$, provided that there is no selection, mutation, migration, nonrandom mating, or *drift*.

H-W$_2$: If an *infinite* population exists with two alleles, A_1 and A_2, with frequencies p and q respectively, then in a single generation the population will settle into genic and genotypic equilibrium with gene frequencies p and q and genotypic frequencies $A_1A_1 = p^2$, $A_1A_2 = 2pq$, and $A_2A_2 = q^2$, provided that there is no selection, mutation, migration, or nonrandom mating.

We will consider selection, mutation, migration, and nonrandom mating to be evolutionary *forces*. One of us has given a detailed argument for doing this (Brandon 2006). The basic argument is that each can be considered a vector quantity and that each quantity can be measured with a common metric (e.g., genotypic frequencies). In contrast, because drift is not directional, it cannot be considered an evolutionary force. This is not simply because it is probabilistic. Selection is probabilistic but has a directional effect.[4] The direction in which drift takes a population can be determined only after the fact, and this is no mere epistemological limitation on our part: if drift is real, this is the nature of that reality (Brandon and Carson 1996).

Now consider H-W$_1$. It is problematic as a zero-force law because it mixes genuine evolutionary forces—selection, mutation, migration, and nonrandom mating—with a nonforce, *drift*. Further, by excluding drift it concludes, quite incorrectly, that stasis is the null expectation. In that, it is entirely misleading even if, strictly speaking, true. It is true in that stasis is, in every generation, the most probable outcome, just as zero is the most probable position of a particle in our model after four time steps. But it is misleading in that most populations will change, just as most particles in our model will end up somewhere other than zero.

What about H-W$_2$? On the face of it, it does not inappropriately mix evolutionary forces with nonforces (because it leaves out drift). But, on the face of it, it applies only to *infinite* populations, and there are none of those. Of course, in science we regularly use idealizations, like that of a frictionless plane, to guide theoretical predictions, and so it is not a sufficient criticism of H-W$_2$ to say that it invokes an idealization. But when, in Newtonian physics, we use the idealization of a frictionless plane to make a prediction about the behavior of a ball, we want ultimately to apply that prediction to a real ball rolling down a real plane where friction does apply. So how do we apply H-W$_2$ to real (i.e., finite) populations? One answer is that we simply stick drift back into the list of evolutionary forces, thus reverting back to H-W$_1$. Then our earlier criticisms apply.

A second answer is that we take infinite populations to be good approximations of real finite populations, thus leaving H-W$_2$ as is and taking its predictions seriously. Then it says: If no evolutionary forces act on a population, its gene and genotypic frequencies will remain unchanged across generational time. That is logically equivalent to the following: If a population does change in gene or genotypic frequencies across generational time, some evolutionary force has acted on it. (Assume, for now, that we are right in categorizing selection et al. as evolutionary forces and drift as not a force.) But we know that these two statements are false! Not only does our basic theory of genetic drift show them to be false, but also all of our broad-based experience with laboratory and natural populations of plants and animals shows them to be false. Virtually all work in molecular evolution is predicated on the falsity of these statements (see, e.g., Bamshad and Wooding 2003; Kreitman 2000; Yang and Bielawski 2000). Best practices in the maintenance of pure strains of laboratory animals are likewise predicated on that falsity. Similarly, perhaps the leading theory of speciation is based on geographical isolation and drifting differentiation, as opposed to selection-based differentiation (Mayr 1963). We could go on. Suffice it to say that the second interpretation of H-W$_2$ is flat-out false.

H-W$_1$ is not, strictly speaking, false, but it is certainly misleading and it certainly does not provide appropriate null hypotheses in evolutionary scenarios. Its evolutionary narrowness—Beatty's point—is another problem, but not one on which we have concentrated. The PD succeeds where the H-W law fails. It is universal and it does provide appropriate null hypotheses for evolving populations: change.[5]

Forces and Null Expectations: Objectivism versus Conventionalism

We have articulated a general, universal law of evolution and have termed it the zero-force evolutionary law. The question we want to address here is this: are there objective matters of fact that settle what count as forces in a particular science, and so what counts as the zero-force condition, or is this a matter of how we set out our theory, and so a matter of convention? (One could also put this question in terms of null hypotheses, but let us stick with this first formulation.)

We will not dare to try to answer this question in general, though we will share our suspicions: in some cases objective facts will settle the matter, but in most cases they will not.[6] But in the present case it is clear that we must take a conventionalist stance (*sensu* Reichenbach 1938). What counts as a zero-force condition for us depends on our choice of how to characterize an evolutionary system. We have chosen a quite minimal characterization, namely any system in which there is reproduction with heritable variation. We think that there are good reasons for this choice—in other words, that it is not an arbitrary choice. But it is a choice and there are alternative ways of theorizing. Let us briefly review our reasons.

First, we choose to look only at reproducing systems because we think that reproduction is central to biology, at least as biology is conventionally understood. Second, variation is nearly inevitable in any system complex enough to reproduce itself. Thus, wherever we find living systems we expect variation. This, we think, is fairly obvious. Finally, and this is not obvious, some degree of heritability is nearly inevitable as well. Not only might some think that this is not obvious, but some might think it false. For instance, Newman and Müller (2000) have argued that accurate inheritance (what they call the "Mendelian World") is an evolutionary achievement, the result of natural selection, and is not evolutionarily primitive (see also Callebaut, Müller, and Newman 2007). We agree. But heritability, in the evolutionarily relevant sense,[7] does not require anything like what Newman and Müller have in mind. As Griesemer (2000) has emphasized, biological reproduction involves material transfer; that is, the parent transfers not simply information, not just a "blueprint," but an actual bit of matter that used to be parent and that now becomes offspring. That is how biological reproduction works. And this material transfer ensures some degree, even if low, of fidelity of reproduction.[8]

Why Not Include Natural Selection? We did not include selection among the basic features of evolving life in the zero-force condition. Why not? After all, we are looking for the conditions that are generic for life no matter where or when it is found. We think that the ZFEL is a feature of universal biology. And we also think that natural selection is an expectable feature of life; that is, we would expect to find natural selection operating almost whenever and wherever we find life. So why not build it into our very characterization of an evolutionary system? We have built variation and heredity into that characterization because of their genericness, or expectability, so to be consistent, should we not also build in natural selection?

Our reason for not doing so is that we wish to keep as an open empirical question the importance of natural selection as a force in evolutionary change. Our goal is to create a framework that better enables us to empirically investigate when and where and how natural selection acts and

interacts with other evolutionary forces. It might seem that we are somehow minimizing the role of natural selection by leaving it out of our
basic characterization of an evolutionary system. On the contrary, we
are convinced that natural selection is of enormous importance in evolution. What the zero-force condition does is give us a neutral background
against which to see selection in action. By analogy, Newton certainly
did not mean to downplay the importance of gravity by stating the law
of inertia as he did. Rather, that law provided the background against
which the role of gravity could be rigorously investigated.

An Apparent Anomaly Explained

Our claim about the ZFEL may seem to be in direct conflict with population genetic theory, in a way that we alluded to in chapter 3. One of
the standard predictions of the theory of genetic drift is that it eliminates
genetic variation from populations. The dynamic of this is easy to understand. In figure 6.2 we relabeled the x-axis of figure 6.1 to be the relative
frequency of allele A_1. But that axis differs importantly from the one in
figure 6.1 in that it has definite endpoints of 0 and 1. The relative frequency of A_1 cannot go beyond 1 or 0. Furthermore, such boundaries are
absorbing boundaries (if we ignore back-mutations), in the sense that,
once the population gets to one of those values, it is stuck there. Given
a random walk with absorbing boundaries the expectation is that each
particle (each population) will eventually move to one of the boundaries.
Thus, according to this bit of theory, drift eliminates genetic variation
from natural populations.

On the other hand, we have said that drift is a source of variation
for the ZFEL, which seems to contradict population genetic theory. And
indeed it does. Fortunately, we are right and that bit of simplistic theory is
wrong. First, as we argued in chapter 3, it is not drift *simpliciter* that acts
to reduce variation but drift plus the boundaries, which are constraints.
Second, as already mentioned, the boundaries are not really absorbing
in the strict sense, because in real populations mutations (not to mention
migration) are always occurring. So the simplifying assumption of absorbing boundaries is never really true. The real question is whether or not the
primary effect of drifting into the boundaries overwhelms whatever it
is that mutation is doing. But this leads into our more important third
point.

Recall our hierarchical approach to drift. Early in the history of genetics allelic differences were determined solely by phenotypic differences. What counted as a single allele from that point of view was, we

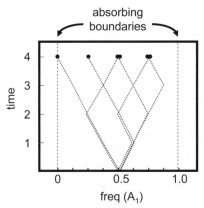

FIGURE 6.2 Multiple drifting populations showing frequencies of allele A_1, with absorbing boundaries at 0 and 1. Unlike figure 6.1, each particle is now a population. Eventually all populations will drift to those boundaries.

now know, molecularly quite heterogeneous. Then came gel electrophoresis. Allelic differences could now be identified in terms of protein behavior in charged gels (a proxy for three-dimensional shape and electrical charge). Again, allelic identity hid a lot of molecular diversity. Now we can sequence strands of DNA and so, if we wish, can make sequence identity the criterion of allelic identity. So what is an allele from the point of view of population genetics? It is a theoretical entity with no fixed molecular interpretation. When a population geneticist says that a population is fixed for allele A_1 by the process of drift, we say that the ZFEL tendencies are working at various molecular levels and that if one were to check at a fine enough scale one would find that A_1 is in fact A_7, A_4, A_{13}, A_{37}, . . . This is where it is important to keep in mind our hierarchical theory of drift. Drift is not occurring at only a single level.

Thus, on the question of populations drifting to fixation, population genetic theory is based on a fiction. Our view conflicts with it. We are not concerned.

Drift as a Causal Concept

The question arises whether drift as we understand it can play the role of cause in evolutionary explanations. Here we argue that it can, giving two very different arguments for that conclusion.

Drift and Probabilistic Explanation. Philosophers have not been able to reach a consensus concerning the nature of scientific explanation. There are many competing theories, and it is well beyond the scope of this book to try to shed light on any of them. However, there is one controversy in the theory of explanation that is relevant to our topic and so we will briefly touch upon it here. The issue belongs in the larger domain of probabilistic causation and probabilistic explanation. Until fairly recently many philosophers thought of causation in exclusively deterministic terms. If the cause was present, then so too would be its effect. In addition, many, but by no means all, philosophers thought of scientific explanation as largely causal. We are ourselves sympathetic to what Salmon (1984) calls the causal-mechanical model of scientific explanation. According to this model, to explain some phenomenon is to explicate the causal mechanism that produced it. If you were to put these two independent beliefs together, you would come to the conclusion that only causally determined events could be scientifically explained.

Many contemporary philosophers find this conclusion unsatisfactory. One way to avoid it is to drop the causal-mechanical model. In our view a more satisfactory alternative is to develop an adequate account of probabilistic causation that would ground such probabilistic explanations. It is not our goal to do that here, but enough progress has been made on that front (see, e.g., Woodward 2003) that we feel comfortable in sketching an account of drift as a causal concept. We do so in full recognition that some will think that you explain the evolution of a trait when you can show how and why it evolved by natural selection. If, on the other hand, you say that a trait has the frequency distribution that it has in a population because of drift, some will say that is no explanation at all. It is against that intuition that we argue.

Fitness is a probabilistic propensity (Brandon 1978; Mills and Beatty 1979; Richardson and Burian 1992). Selection is a probabilistic sampling process. Natural selection, which we all think of as explanatory, is just this sampling process playing itself out (largely) according to the probabilistic expectations. Drift is not a different process. It is exactly the same—that is, a probabilistic sampling process. However, drift occurs when things do not work out according to the probabilistic expectations. If we can explain one event (the natural-selection event) by subsuming it under the causal process that produced it, then we can explain the other (the drift event) by subsuming it under the very same process. This is exactly parallel to Salmon's (1971, 1984) example of radioactive decay. He argued that, if you can explain some highly likely event, say the decay of a polonium-218 atom during a one-hour time period (the half-life of polonium-218 is 3.05 minutes), you can also explain a highly unlikely event—the nondecay of a polonium-218 atom during a one-hour time

period. In both cases the explanans is the same. The outer band of elec-
trons is in an excited state with a certain quantifiable propensity for decay. According to quantum theory, that is all there is to it. No other information is relevant.

Some might think of drift as the absence of cause. But, as we see it,
the relevant causal understanding is the full set of objective probabilities
that govern the entities to be sampled. Sometimes the probable happens;
in fact, usually the probable happens. But sometimes the improbable
happens. In either case, causal understanding is achieved when we as-
semble the relevant probabilities governing the events in question.

A Newtonian Analogy. We can make the same point about drift being
causal in a very different way, using our Newtonian analogy. Newton's
first law sets the default state of Newtonian objects. This is what they do
if nothing happens to them, if no net force impinges on them. Newton's
second law, $F = Ma$, then gives the means to do quantitative dynamics.
It, of course, applies to any Newtonian force. Newton's gravitational
law, the inverse square law, describes the behavior of one particular
force. Consider a paradigmatic Newtonian explanation for, say, the fall
of an apple (in a vacuum—it is much easier). To a first approximation,
we measure the mass of the Earth (M_1), the mass of the apple (M_2), and
the distance between their centers of gravity (d); depend on someone
else to give us the gravitational constant (G);[9] and then plug all of this
into the gravitational law to get the force exerted on the apple. We then
use the second law to get the acceleration of the apple. That explains
the trajectory of the apple as it falls to the ground. By any philosophical
account of scientific explanation, this is a good scientific explanation.
But it also seems to be a good causal explanation. If we think of Newtonian forces as Newtonian causes, and that seems natural, then we have
cited a cause (gravity) and thereby explained the event (the acceleration
of the apple).

Now consider another Newtonian phenomenon: an apple at rest on
the ground. Although forces are acting on it, no net forces are; thus, $F = 0$.
According to the second law, a should equal 0 as well. And it does. This
apple is obeying Newton's first law. Have we explained its behavior? It
certainly seems so. We have subsumed it under Newton's first and second
laws. But unlike the first case, no net force is acting on it, and so one might
think that this explanation is noncausal. But we think it more reasonable
to say that it is causal, that Newton's laws describe the causal structure of a Newtonian world and that the apple is behaving accordingly.
But, unlike the first case, which we will call a *special causal explanation*

because it cites a special cause, this case simply cites the absence of special causes and so relies on the default state, here the law of inertia. We will call this sort of explanation a *default causal explanation* (Brandon 2006).

The argument here is that insofar as the Newtonian explanation of inertial phenomena is causal, so too are our explanations of ZFEL phenomena.

Probability Theory as the Reductive Foundation for All of Evolutionary Theory

The focus of this book is not on selection. That is not because we think selection is unimportant, but rather because we think its importance cannot be properly appreciated except in the context of the ZFEL. We have argued that the ZFEL is a fundamental law of universal biology. However, we also believe that the principle of natural selection (PNS) is such a law. One of us (RB) has for some time argued that the PNS is a particular instantiation of the law of likelihood from probability theory. Consider this statement of the PNS:

> If *A* is better adapted than *B* in environment *E*, then (probably) *A* will have more offspring than *B* in *E*.

If we spell out relative adaptedness in terms of probabilistic propensities, as Brandon has argued for, then this is simply an application of the principle in probability theory that allows one to infer from probabilities to frequencies (the principle of direct inference). Is such a principle an analytic statement, a bit of pure mathematics? Brandon and Rosenberg (2003) have argued that it is not, that this is exactly the lesson of Hume's critique of induction. As Hume showed, it is not analytic that the sun shall rise in the east tomorrow morning, nor that bread shall nourish us tomorrow. Similarly, although it is true that in the past the probable has happened more frequently than the improbable (that is how we define the probable, by means of the principle of indirect inference, or inference to chance), it is not analytic that this shall continue. All hell may break loose tomorrow.

Thus, according to this view, the PNS is a deep synthetic truth about the world that has a special application to biology. Not just any application of the principle of direct inference is relevant—the PNS is structured in ways that makes it biological (compare reproduction within common selective environments).[10] But it is ultimately reducible to a bit of probability theory, not the probability calculus but the bit that allows the calculus to be applied to the empirical world.

Now it should be fairly clear that the ZFEL is also ultimately reducible to probability theory. Let us distinguish the arguments we have offered for the application of the ZFEL to particular biological phenomena from the arguments we have offered for the basic truth of it. The latter have been abstract and based on facts about random variation and how ensembles of randomly varying entities behave. We have been dealing with substrate-neutral sampling processes, so it should not be surprising that physicochemical reduction is not at all plausible here. Rather, for the ZFEL, as for the PNS, probability theory seems to provide the reductive foundation for a universal evolutionary theory.

This is interesting. Why? Because reductionistically minded philosophers have never even envisioned anything like this—"this" being the reduction of an empirical science to what might seem to be a branch of pure mathematics. But, of course, we maintain that there is a bit of probability theory that is not pure math. So perhaps this is where we commit our philosophical heresy.

The further question arises whether one should think of the ZFEL as analytic or synthetic. It might seem that it is analytic. After all, the ZFEL follows deductively from probability theory. If we accept that probability theory is true, then given a system with reproduction and heredity, diversity and complexity *must* increase. The ZFEL *must* be true, analytically. However, if probability theory is synthetic, then the ZFEL must be synthetic as well. On the other hand, it is not currently practical to test probability theory, and therefore, for practical purposes we can treat the ZFEL as analytic, in which case—a biologist might wonder—is there really any need to test it? We think the answer is yes, if only because it is always important to confirm that our logic is sound.

The ZFEL and the Second Law of Thermodynamics

A more normal course for the reductionist with respect to the ZFEL (but probably not the PNS) is to look to the second law of thermodynamics. The ZFEL essentially describes a diffusion process, and the second law is widely acknowledged to govern such processes in physical systems. Or at least the statistical-mechanical interpretation of the law is used in this way. For example, the increase in diversity of the pickets in the picket fence described in chapter 1 would ordinarily be said to be a consequence of the second law.

Some have taken this route in biology. Some decades ago, a number of biologists invoked the statistical-mechanical interpretation of the second law to explain the increase in diversity of life in a way that is reminiscent of our treatment of the ZFEL for diversity (Brooks and Wiley 1988). Their argument was not very clearly articulated, making it difficult to say how much overlap there really is with ours. At a minimum, it would seem that our view is much broader, in that ours very explicitly extends to the internal complexity of organisms, understood as the degree of differentiation among their parts, in a way that these earlier treatments did not. Also, it is clear that these earlier treatments found much of their vocabulary and inspiration in physics, that a significant part of their mission was to forge a link between physics and biology. Consistent with this, they invoked the second law as foundational. We do not. And consistent with our decision, we identify the intellectual ancestor of our view as Herbert Spencer rather than Ludwig Boltzmann.[11] But why? Why do we not invoke the second law?

The first reason is simply that we did not need to look there. We never needed to invoke the second law to explain anything that we did not already have a simpler and more general explanation for in probability theory. Finding sufficient basis for the ZFEL there, we have had no need to turn to physics.[12] Of course, given the similarity of underlying principles, we cannot help but speculate that the second law itself might ultimately have its basis in probability theory as well. Time will tell. But given that probability theory is more general than thermodynamics, there is no need for us to wait.

Second, we point out the following disanalogy between the ZFEL and the second law. Entropy, or disorder, has a common measure in physics.[13] As a consequence of that, it is impossible for a region of spacetime to increase in entropy (in accordance with the second law) while every subpart of that region decreases in entropy. Entropy of the whole is reducible to entropy of the parts. Of course, some part can increase in order, but this has to be at the expense of disorder in adjacent parts. Diversity and complexity, as we have defined them, do not work this way. The reason they do not is that both are hierarchically defined, and the measure of diversity/complexity at one level is not commensurable with measures at adjacent levels. For instance, organismic complexity may be measured in terms of number of cell types (or, more generally, degree of differentiation among cells). But there is no conceptual dissonance in saying that a complex organism (one with more cell types) is composed of simpler cells than a less complex organism (one with fewer cell types). Indeed, McShea (2002) found just this. Our point here does not depend on whether or not that finding is correct. The point is that it is a physical possibility. And the reason it is, the reason it is not contradictory to say that the whole achieves greater complexity while reducing the complexity of its parts, is that cell complexity has an entirely different measure than organismic complexity. Cell complexity might be measured in terms of number of different types of subcellular structures, such as organelles. But organismal complexity, as we have seen, can be measured at a different level, in terms of number of different types of cells. Similar remarks could be made about diversity. Thus, the ZFEL is, in this way, fundamentally different from the second law of thermodynamics.

A Generalized ZFEL for Physical Systems

The diversifying and complexifying picket fence in chapter 1 might seem like a fair analogy for the increase in variance that lies at the heart of the ZFEL. Essentially each picket is sampling the distribution of possible accidents, and the pickets come to differ from each other when that sampling produces different results in each picket. But the analogy is imperfect, because—although it is helpful for conveying the idea of accumulation of variation—there are two big differences between biological systems and physical systems like this. Picket fences do not meet two of Darwin's three conditions. There is no heredity (condition 2), in the sense of offspring resembling parents, and there is no differential reproduction (condition 3), and the reason in both cases is the same: there is no reproduction at all. The ZFEL is intended as a biological principle, and if reproduction and heredity are basic to biology, as we argued briefly above, then technically the ZFEL does not apply to the picket fence.

However, one can imagine a more general version of the ZFEL, one that applies equally to biological and nonbiological systems, in which reproduction is understood to be a special case of something like "persistence" (Bouchard 2008) and heredity is understood to be a special case of something like "memory." In this more general understanding, reproduction would be just one route to persistence, the route biology employs in a world of mortal organisms. It is a mechanism that increases the probability that a given phenotype in existence at some time will also be present at some later time. The organisms die but the lineage persists. And inheritance in biology is just the property of organisms that accounts for the persistence of the original phenotype and also the persistence of any variations that have arisen.

Returning to the picket fence, the pickets do not reproduce but they do persist, more or less intact, with the passage of time, with the result that each picket at some later time is very similar to the same picket at some earlier time. And there is no inheritance, but there is retention of variations, memory, so that variations arising in any given picket at some time will tend to be present at some later time as well. With the concepts of reproduction and inheritance broadened in this way, it is clear that one could develop a correspondingly general version of the ZFEL—the G-ZFEL—that would look something like this:

> G-ZFEL (special formulation): In any system in which there is persistence, variation, and memory, in the absence of forces and constraints acting on diversity and complexity, diversity and complexity will increase on average.

There would be a more general version as well:

> G-ZFEL (general formulation): In any system in which there is persistence, variation, and memory, there is a tendency for diversity and complexity to increase, one that is always present but may be opposed or augmented by forces or constraints.

However, developing and arguing for such a principle is beyond the scope of this book.[14]

7

Implications

So far we have been concerned mostly with explicating the ZFEL. In this last chapter, we discuss some of its implications for biology.

Diversity

Diversity Is Easy, Stasis Is Hard. This is the lesson of the ZFEL. The third position in codons, largely removed from selection, changes within populations and across them. Nonfunctional intergenic DNA diverges among populations, species, and higher taxic units in a way that provides reliable information about times of divergence of these units. In addition, genes that are functional from a molecular point of view, but not necessarily from an ecological point of view, also diverge between isolated taxic units. Thus, for instance, selection for reproductive isolation is not required for allopatric speciation, because geographically isolated populations will tend to diverge from one another without any input from selection.[1] In macroevolutionary studies, where convergent and parallel evolution require special explanations, random ZFEL-like divergence is the recognized null expectation. Finally, diversity increases and diversity decreases are not symmetrical. The ZFEL tendency is for diversity to increase. Thus, although selection may favor diversity increase, selection is not needed to achieve it. On the other hand, it is less likely that an evolutionary system will spontaneously decrease in diversity. Wherever multiple extinctions occur, exogenous causes are routinely sought. And the ZFEL adds that diversity stasis—some extended period of time when the diversity of some group or some ecological unit remains constant—presents the same puzzle, requiring the invocation of constraints or forces sufficient to overcome the ZFEL tendency.

Diversity: A New View of Change. With the exception of the final point—that diversity stasis and diversity decreases should be treated differently from diversity increases—there are no novel observations in the last section. But general evolutionary thinking has not caught up with what is already known about the ubiquity of ZFEL-governed diversification. More broadly, it has not grasped that a new view is implicit in this way of thinking, a new view not just of diversification but of *change*, one that goes beyond the ZFEL. We explain in the four subsections that follow.

Response to selection. Consider first the fundamental equation in quantitative genetics: $R = h^2S$ (where R is the response to selection, h^2 is the heritability of the traits under selection, and S is the selection differential). Technically, this formula is true, but it can be misleading in that the technical meaning of R is much narrower than its nontechnical meaning.

Quantitative genetics deals with traits, like height, that vary continuously rather than discretely and that are influenced by multiple genetic loci. The basic descriptive vocabulary of quantitative genetics includes the mean and variance of the distribution of some quantitative trait. R is defined as the difference between the means of the distributions of generation 1 and generation 2. But consider cases of stabilizing selection. Suppose we are dog breeders and have found a phenotypic value of some trait that we prize. Suppose further that this value is the mean value in the population, but that there is significant variance about this mean. We strongly select for that value, letting only those at or very close to that value reproduce. What happens? Our expectation is that the mean will remain unchanged but that the variance will decrease. This is a response to selection; that is, the distribution of trait values has changed due to selection, but technically $R = 0$ because the mean has not changed. But now suppose we have reached a mutation-selection equilibrium such that the continued imposition of our artificial-selection regime no longer decreases the variance but only maintains the mean. Selection is still having an effect in that the mean and variance of the population are different than they would have been had selection been removed. But, again misleadingly, $R = 0$.

So what? Why not just conclude that the breeders' formula applies only to cases of directional selection and leave it at that? The reason we bring this example up is to show how a failure to recognize the ZFEL has affected our basic evolutionary concepts. Remember that the Hardy-Weinberg law leads to the view that "evolutionary change = action of evolutionary forces." This in turn leads to an overemphasis on directional selection and an underappreciation of the necessity of the action of evolutionary forces for evolutionary stability. The principle of drift, discussed in chapter 6, says that a drifting mean is the natural state of an evolutionary system. Our ZFEL view says that diffusing variances is also the natural state. The point is that, contrary to standard intuitions in evolutionary thought, stability is a signal of strong evolutionary forces.

"Stasis is data" (Gould and Eldredge 1993). We offer one example from the recent history of evolutionary biology to support our contention about standard intuitions. When Eldredge and Gould (1972) introduced the theory of punctuated equilibrium, it was taken to be quite controversial among most microevolutionary biologists. The primary reason for this is that Gould suggested that the pattern of punctuated equilibrium requires some process not covered by standard population genetic theory. Critics immediately focused on the punctuation periods, that is, the periods of rapid evolutionary change that from a paleontological viewpoint were brief. Most argued that no new evolutionary process was required to explain such episodes, since from a population genetic point of view these brief bursts of evolutionary activity occurred over thousands of generations, which was ample time for natural selection to produce evolutionary change. To our knowledge, none of these microevolutionary critics focused on the long periods of stasis, both in means and in variances. However, from the ZFEL point of view, it is the long periods of stasis that are remarkable. That is, if the paleontological pattern that Gould and Eldredge pointed out is robust, the phenomena that really need explanation are the long periods of stasis. We think this can be done, with a large input from population genetic theory, but only against the proper backdrop: drifting means and ZFEL-driven increase in variance.

If professional evolutionary biologists think this way, it is unsurprising that the lay public does too. Clearly, religiously based creationists see the need to invoke an external cause, a creator, to explain evolutionary change. But we are more concerned with the scientifically minded lay public, for instance, the sort of undergraduates we teach. How many times has the question arisen in undergraduate classrooms: is human evolution over? The presupposition behind that question is, first, that humans are no longer under selection and, second, that the absence of any selective force means that no evolution is occurring. Typically, the discussion centers on the first issue, whether or not natural selection is occurring. And that is an interesting question. Our point is that settling it will not answer the original question, whether or not humans are still evolving. Because even if selection were completely absent in modern human populations, we should expect evolutionary change, both in means (arising from the principle of drift) and in variances (driven by the ZFEL).

At least some professional philosophers take the standard view as well. For instance, in his highly influential account of causal explanation, Woodward (2003) assumes that the persistence of traits in a lineage requires no special causal explanation, that instead such phenomena are analogous to the pervasive background-radiation signal of the big bang. The background radiation was causally produced by the big bang and therefore is explained by it, but we do not need to continually explain its persistence. In other words, Woodward assumes that the background radiation needs no special auxiliary interventions in order to persist. So it goes, thinks Woodward, for the persistence of ancestral traits. Here we might think of the genetic code or, on a shorter timescale, the persistence of the horseshoe crab's gross skeletal morphology, largely unchanged over 400 million years. But if, as we have argued, change is the default condition of evolutionary systems, then such persistence requires special causal explanation (presumably in terms of constraints or stabilizing selection or both).

Types of evolutionary equilibria. The widespread misunderstanding of evolutionary stasis discussed above has led to important technical misconceptions of evolutionary equilibria in the levels of selection literature that lies at the intersection of philosophy of biology and theoretical population genetics. In population genetics, at least three sorts of equilibria are usually recognized: stable, unstable, and neutral. Hardy (1908) described the sort of neutral equilibrium that the Hardy-Weinberg law applies to in the absence of evolutionary forces (R. C. Lewontin, pers. comm., 2005). Neutral equilibria are transient. Indeed, strictly speaking, they should last only a single generation (with drift changing gene frequencies each generation and producing a new Hardy-Weinberg "equilibrium"). This is the equilibrium of a small ball on a large, frictionless table. In the absence of perturbations, the ball does not move. This is what is conventionally understood as a neutral equilibrium. We would point out that, in this sort of situation, the word "equilibrium" is misleading. Perturbations are ubiquitous in this world, so in fact the ball is in constant motion. Hardy used the word "stable" to describe this sort of equilibrium in populations but wisely put that word in scare quotes, recognizing that neutral equilibria of this sort really are not stable.

The distinction between unstable and (truly) stable equilibria will be familiar to most readers. In the first case, perturbations from the equilibrium point lead to a dramatic departure from equilibrium (sometimes to a new equilibrium point, sometimes not), as in a ball balanced on the edge of a bowl. While in the second case, there are forces that tend to restore the population to its original equilibrium point when it is perturbed, as in a ball sitting at the bottom of a bowl. It is the latter that might be thought of as properly modeling evolutionary stability. However, genic selectionists systematically misdescribe equilibria actively maintained by selection (Brandon and Nijhout 2006).[2] They describe a bizarre hybrid sort of state in which selection is absent precisely at the stable equilibrium point but in which departures from equilibrium generate (frequency-dependent) selection to restore equilibrium. It is as though they were imagining that a ball resting at the bottom of a bowl has no forces acting on it but that forces appear the moment the ball starts to move. Consider a genotypic model of overdominance, for example, the familiar case of sickle-cell anemia. Here the two homozygotes are strongly selected against (because the wild-type homozygote is susceptible to malaria and the sickle-allele homozygote produces a malformed hemoglobin molecule, leading to the anemia that gives the condition its name), while the heterozygote is selectively favored. This situation quickly leads to an equilibrium (the value of which depends on the relation between the selection coefficients associated with each homozygote). The genic selectionist has no selection acting at equilibrium, and selection differentials increase (usually slowly) as the population is perturbed away from equilibrium. But a better description would say that selection acts in a frequency-independent way, so that at equilibrium the selection differentials between the heterozygote and the two homozygotes are as strong as they are anywhere else in state space. This is what Brandon and Nijhout call an equilibrium actively maintained by selection. It is the equilibrium of a ball at the bottom of a bowl, its stable position maintained by the force of gravity and a counterforce applied by the bowl beneath it. Brandon and Nijhout show that the evolutionary trajectories predicted by these two descriptions are not empirically equivalent, because drift affects them differently. In particular, the two models do not agree on the regions of state space where selection dominates drift and vice versa. The result is a refutation of the widely held view that genic and genotypic models of selection are mere verbal variations of the same phenomena and so one is free to use either or both.

Fitness and evolution. So pervasive is the view that "evolutionary change = action of an evolutionary force" that even our most basic evolutionary concepts are infected with it. Consider the concept of evolution. Population geneticists usually define evolution as *change* in gene frequencies over generational time. From the quantitative genetic point of view, evolution is *change* in the mean of a phenotypic distribution over generational time. A somewhat broader version of this view would say that evolution is *change* in the mean or variance (or some other statistic) of a phenotypic distribution over generational time. Evolutionary-developmental biologists may want to define evolution as *change* in development over generational time. It should be clear that these definitions are compatible with one another, and from our point of view their similarity to one another is more interesting than their differences. They all describe evolution in terms of change.

Of course, the meaning of the term "evolution" is deeply entrenched in biological discourse. So we are probably powerless to fundamentally alter it. However, the entrenched meaning does have strange consequences. In the cases of stabilizing selection discussed above, it is clear that evolutionary forces (primarily selection) are in action. They have a transgenerational effect; in other words, the phenotypic distribution is different than it would have been in the absence of selection. But, if stabilizing selection has produced an equilibrium, then no change is occurring, and therefore, according to the entrenched meaning, no evolution is occurring. So in the end we are forced to say that in these cases evolutionary forces are having transgenerational (it is tempting here to say evolutionary) effects, yet there is no evolution. And that sounds odd.

This oddity could easily be avoided by adopting a more appropriate concept of evolution. We would define evolution as the transgenerational transition in state space of a population (where "population" is understood broadly and hierarchically). For instance, in a population genetic framework, the state space might be described in terms of allele frequencies and their combinations, that is, genotypic frequencies. A transgenerational transition may take the population to exactly the same position in state space that it had in the previous generation, or it may take it to a different position. Neither sort of transition would be privileged in our conception of evolution, because from the ZFEL point of view, stasis is evolution too.

Finally, the privileging of change, the failure to see the fundamentally different causes of change (default vs. special), and the failure to see stasis as a strong signal of selection together have led to empirically inadequate and explanatorily empty conceptions of fitness.[3] Thus, starting with Fisher (1930), many have sought to define fitness in terms of evolutionary *change*. (For Fisher, fitness is the per capita rate of increase of a type.) Recently, Matthen and Ariew (2002) have argued that evolutionary biologists *must* measure fitness in terms of evolutionary change. But if one of the primary ways selection acts is in stabilizing selection, then these views are entirely wrongheaded. And in fact, biologists measure fitness in terms of evolutionary change only very rarely and only as a last resort (Endler 1986; Kingsolver et al. 2001; Brandon and Ramsey 2007).

Complexity

Complexity Is Easy, Simplicity Is Hard. The ZFEL says that in the absence of selection and constraints, random variation will tend to make any set of parts in an organism more different from each other. It is therefore no surprise that unselected genes, pseudogenes, spontaneously vary to produce genome differentiation. Nor is it surprising that in a vertebral column, mutation leads spontaneously to differentiation among the vertebrae, to new part types; nor that in morphology generally, random variation leads to differentiation of the left and right sides of the body. Organisms are highly redundant systems, and unless opposed, random variation will tend spontaneously to break down that redundancy, producing differentiation. Complexity is expected. Complexity is easy.

If complexity is easy, seemingly it should increase over the history of life, and at least impressionistically, it does—but not always and not everywhere, not as universally as the special formulation of the ZFEL predicts. Simple organisms persist, and decreases in complexity—as in the evolution of some parasites—are common. How to explain this? Conventionally, decrease is explained by selection, in the case of parasites by the selective advantages of a streamlined morphology and physiology. The ZFEL says this sort of explanation is apt. Life, for the most part, does not exist in the zero-force condition. Selection is ubiquitous. But the ZFEL also warns that the simplifying variation required for selection to act may be available to selection only rarely (unless developmental organization happens to favor it), and therefore, selection for simplicity may have to await the rare variant. If so, it implies that, when complexity reduction does occur in evolution, it is hard won. Simplicity is hard.

By the same reasoning, complexity stasis is also hard. Complexity stasis falls within the more general domain of morphological stasis. And the conventional view is that stasis requires stabilizing selection or constraints, because otherwise morphology will tend to drift. The ZFEL adds that morphology will also tend to complexify, that the variation among parts will tend to rise. And when it does not, when there is stasis, some external force or constraint is needed to account for it.

Colloquial Complexity Again. The colloquial notion of complexity is deeply culturally embedded. And so we expect that one response to the ZFEL will be: "That's all very well. The ZFEL may be a source of nonfunctional genes and nonfunctional parts. But that's not complexity." Meaning, of course, that it is not—in our terms—colloquial complexity. It is not some ineffable combination of concepts like part types, emergence, functionality, integration, sophistication, and so on. In essence, the complaint is that the ZFEL does not produce structures like image-forming eyes and the mammalian brain. And that is true. It does not. Nevertheless, the ZFEL may be relevant to the evolution of these structures. Whatever colloquial complexity is—and we have doubts about whether it can be defined well enough to make it useful in biology—pure complexity is probably a rich source of raw materials for the production of it. Just as nonfunctional genes, junk genes, are a rich source of raw materials for genomic complexity in the colloquial sense, junk parts may be a rich source of raw materials for colloquially complex structures like eyes and brains. Who is to say that one of the various vertebral deformities Ehling observed could not be the foundation of the next adaptive innovation in vertebral-column design? Thus, the ZFEL does not give us colloquial complexity. Only selection can do that. But it does say that, both in genes and in morphology, the raw materials for colloquial complexity are expected to be enormously abundant.

This argument might not seem especially new. Darwin ([1859] 1964) pointed out that duplicate parts offer opportunities for the evolution of adaptive novelty, roughly what we are calling colloquial complexity. And the evolutionary opportunities for colloquial complexity that are offered by duplication and differentiation have been well explored since then, especially recently, and especially in molecular biology (Lynch 2007a, 2007b). But the ZFEL makes a different point, saying both more and less than that. It says less in that what it predicts is only parts, not functional parts. It predicts novelty but not adaptive novelty, pure complexity but not adaptive complexity. But the ZFEL also says more. It identifies more than just opportunities; it identifies a *tendency*. And this tendency may make the job of natural selection in the production of colloquial complexity especially easy.

We explain this idea in the next two subsections, but first we need to warn that our arguments do not address colloquial complexity in its fullest sense. What we address is colloquial complexity in the sense of part types *plus some notion of functionality,* hopefully capturing some large fraction of whatever it is that eyes, brains, automobiles, and computers share, of whatever it is that makes people want to call these things complex. We acknowledge that our hybrid notion of parts and functionality is missing a lot. It is missing any notion of difficulty of manufacture, sophistication, emergence, and so on, all components of colloquial complexity in its fullest sense. We also acknowledge that it is inadequate for most scientific purposes. We have no idea how to operationalize it. Still, it is a first step toward making a connection between our pure complexity and the street notion of complexity. And it is clear enough, we hope, to use it to introduce some potentially fruitful speculation.

Colloquial Complexity and the Tinkerer's Assistant. Jacob (1977) compared natural selection to a tinkerer. He argued that natural selection designs organisms, not as an inventor designs a machine, from scratch, but rather by tinkering with existing designs, improving them in small increments. The tinkerer is in some ways an apt metaphor for natural selection,[4] but for complexity the ZFEL says the scenario is incomplete. The ZFEL says that, in the evolutionary workshop, the tinkerer is not alone. There is also an assistant present, an odd character whose main job seems to be to attach parts to the evolving machine, not just parts but novel part types. Occasionally, rarely, the assistant removes a part type, but mostly he adds them. Now most of the new part types the assistant adds have no conceivable function, and the tinkerer rejects them, plucking them off the machine as fast as he can. Sadly, the main effect of the assistant's activity is to discombobulate the machine and to distract the tinkerer, who must spend much of his time removing the clutter of troublesome new parts. Luckily, some of the parts are useless but benign, and the tinkerer can ignore them, allowing them to remain in place. And happily, a very few are actually improvements, and these the tinkerer rapidly incorporates into the machine's design. It turns out that the assistant is an excellent source of raw materials—new ideas, so to speak—for the tinkerer in his quest for improvement.

The analogy invites a speculation. Consider the vertebrate eye. It counts as complex in our pared-down colloquial sense because it has many differentiated parts and because it can produce an undistorted image on a retinal surface; in other words, it functions. Now only natural selection can produce function, and therefore, the evolution of (pared-down) colloquial complexity needs natural selection. That is, it needs the tinkerer. But—and here is the point—colloquial complexity also benefits enormously from the assistant, the ZFEL. The assistant could be a major source of raw material, new part types, new ideas, out of which natural selection can craft colloquially complex designs. Just as we can see partly randomized nonfunctional DNA, pseudogenes, as raw material available to selection for the production of new functional genes, we can also see partly randomized nonfunctional morphological parts as raw material for the evolution of colloquial complexity of morphology. Present data suggest that the DNA of at least some organisms is full of pseudogenes, each one a potential future novel gene. Could it also be that higher-level morphology is also replete with nonfunctional parts?

On the standard assumption that selection should rapidly eliminate nonfunctional parts, the answer would be no. And that answer could be right. It could be that the tinkerer is able to keep up with the assistant, removing benign additions as fast as the assistant can add them. Organisms could be quite streamlined.[5] But cave crayfish and cave fish retain certain eye structures many generations after their eyes have stopped serving any obvious purpose. Some whales still have the remains of hipbones, apparently useless to them, millions of years after they lost their function in hindlimb support. And in humans, it is questionable whether structures like earlobes, appendices, and tailbones have any important function. And many people are born without a certain muscle of the forearm, the palmaris longus, with no apparent effect on their abilities.

How common are useless structures? In his discussion of "rudimentary, atrophied, or aborted organs," Darwin remarked that "organs or parts in this strange condition, bearing the stamp of inutility, are extremely common throughout nature" ([1859] 1964, 418). What is more, many of these useless parts could be structures that—unlike the appendix—have been functionless since the moment of their origin. Consider some externally salient candidates in humans, such as the earlobes, the folds of the navel, the slight webbing between fingers and toes, the folds of flesh associated with obesity, the concavities under the arms, and the variation in pigment over the surface of the body. All of these are differentiations of the skin and therefore count as pure complexity at some level. What is more, we note that these are differentiations *just* of the skin. Presumably, every organ and organ system has its own large complement of functionless structures. The ZFEL raises the possibility that, like pseudogenes, the pure complexity represented by such useless parts could be the raw input to the selective process that generates novel functional parts and combines them to produce colloquial complexity. And that raw input could be extraordinarily diverse and abundant.

The tinkerer story suggests a second speculation. It might seem that, despite the occasional happy discovery of a use for a new part type in a colloquially complex design, the assistant is mainly a problem for the tinkerer. Many of the new part types he cannot use, and he cannot pluck off all of the useless ones fast enough. Some will remain, sometimes for a long time, adding to the general clutter and occasionally disrupting normal operation of the machine. So given the tinkerer's opportunistic mind-set, it would not be surprising if he sometimes opted to make the best of a bad situation, to make use from time to time of parts he did not really need, to pursue more-complex designs in preference to simpler ones, even when the more complex is less than ideal. It would be much easier for him to choose a design that uses one or two of the novel parts coming in than to choose a simpler one, a design that might require him to wait for the rarer loss of the right sort. Thus, when a novel functional design is discovered, it should not be surprising that it is often complex, perhaps more complex than it really needs to be. Let us exit the metaphor. The ZFEL acknowledges that the functionality that makes vertebrate eyes so impressive requires natural selection. But it also points out that, if the flux of novel part types is high enough, then it should come as no surprise that, as eyes evolve, they become more complex in the pure sense. And further, the ZFEL raises the possibility that eyes may be more complex than they really need to be, that simpler designs might have done the job just as well or better.

Colloquial Complexity by Subtraction (and Implications for Intelligent Design). There is a notion—lying near the bottom of the pool of assumptions that scientists and even lay onlookers bring to their evolutionary thinking— that complex structures must arise "by addition." In other words, the assumption is that a structure with many part types must arise by addition of part types to some ancestral structure that was simpler, with each successive addition favored by natural selection. An assumption along these lines is implicit in Darwin's story about the vertebrate eye, in his story about how it could have evolved from a much simpler eye in successive steps. The ZFEL challenges this implicit assumption, raising the possibility that colloquially complex adaptations could also have arisen "by subtraction." We will explain what this means shortly, but let us stress at the outset that the argument is purely speculative. We raise it as a possibility worth considering.

Consider two different ways of building a stone arch. One could build it by addition, starting with two stone piers separated by some distance and adding one stone at a time to each side until the stacks meet in the middle, at the keystone, completing the arch. Doing it this way, each stone must be specially carved, and supports of some kind are needed for each new stone added, until the keystone is in place, at which point the arch will be self-supporting. In evolution, the analogous line of thought is that complex organismal structures are assembled by addition, with parts that are sculpted by selection and added in such a way that they are "supported" by natural selection—in other words, that they are functional—every step of the way. The intermediates must be advantageous (e.g., Dawkins 1986; Lenski et al. 2003).

But there is an alternative method. It begins with a huge pile of irregular stones, none of them specially carved. In a sufficiently large pile, there will be some small subset that naturally form a self-supporting arch of some size smaller than the pile. In other words, somewhere within the huge pile, there is likely to be some set of stones that find themselves, by chance, in a self-supporting arch-shaped configuration. No arch will be visible to an onlooker. But it is there, buried among the many looser stones that do not participate in its support. To see the arch, for it to be revealed as a complex self-supporting structure, the superfluous stones must be stripped away. Exiting the analogy, we raise the possibility that complex adaptive structures arise spontaneously in organisms with excess part types. One could call this self-organization. But it is more accurately described as the consequence of the explosion of combinatorial possibilities that naturally accompanies the interaction of a large diversity of arbitrary part types.[6] If there are enough of them, some subset of them will be functional, for something. And this subset will be maintained, and presumably tuned toward optimality, by natural selection. It should be obvious where we are imagining such a huge excess of part types comes from. It comes from the ZFEL, which invites us to see organisms as always awash in novel part types. The vast majority of these will be nonfunctional, of course, and most of the nonfunctional ones will ultimately be removed by natural selection (even as new part types are arising). But it is their removal that reveals wondrous adaptations we call complex in the colloquial sense. Colloquial complexity is revealed by subtraction. And as for intermediates, by this route, there simply aren't any.

If plausible, this mechanism is relevant to a current controversy in evolutionary and creationist circles. Advocates of what is called "intelligent design" contend that certain structures in organisms are too complex to be explained by natural selection. In our terms, the controversy is about structures that are complex in the colloquial sense, structures that have many part types and are also functional (like the molecular motor of the bacterial flagellum, now the canonical intelligent-design example). The creationist intuition, shared by many lay onlookers to the debate, is that it is difficult to see how the intermediates of these complex structures could have been functional, and therefore how they could have arisen and been maintained by natural selection. It is the arch question, and it takes for granted that evolution must proceed by addition. The standard Darwinian response accepts that assumption and tries to show how the intermediates could indeed have been functional. Further, to make complexity by addition more plausible, evolutionists point out that intermediates may have served functions that were different from the final one we are seeking to explain.[7] We do not doubt that this occurs and often. Evolution by addition combined with change of function, or exaptation, is likely an important route to colloquial complexity. But we point out that there may be another route available as well. If novel part types are delivered in excess, as the ZFEL suggests, then the combinatoric possibilities could be vast, with the result that colloquial complexity could—like pure complexity—be easy. And the role of natural selection could be mainly negative, revealing colloquial complexity by subtraction.

"Order" from Disorder. We have not used the word "order" so far in connection with the ZFEL, because we did not need it and because—like "complexity"—the word has so many different meanings, some mutually contradictory. Order can mean redundancy. A picket fence is said to be ordered on account of the regularity and similarity of the pickets. Order in this sense is the opposite of complexity. But order also refers to a hard-to-specify combination of complexity, regularity, and functional specificity. Organisms are said to be ordered because they contain many parts, and each interacts in a regular way with other parts, such that they are able to perform some function. We think order in this sense has been the source of a great deal of trouble in evolutionary biology. Many people thinking about the origin of life are puzzled by how "order" could have arisen from disorder, how a "disordered" primitive ocean could have given rise to an "ordered" bacterium and ultimately to a hugely "ordered" mammal. The apparent contradiction of the second law implied by order-from-disorder led the physicist Erwin Schrödinger to posit a still-puzzling and physically uncharacterizable quantity, "negentropy," to account for it. And religious creationists use the improbability of "order" arising from disorder to raise doubts about evolution and a naturalistic origin of life.

Poorly constrained concepts create conundrums. "Order" in the sense of pure-complexity-plus-functionality-plus-regularity-of-interaction throws together several concepts that have no necessary connection to each other.[8] But when torn from their illicit embrace, the order-from-disorder problem disappears. The ZFEL accounts for pure complexity. Or to put it another way, for pure complexity there is no complexity-from-disorder problem, because differentiation of parts just *is* disorder. Nor is there a functionality-from-disorder problem. Natural selection takes the pure complexity, the "disorder" handed to it by the ZFEL, and preserves the instances of it that are functional. The rules of interaction among parts follow from a combination of selection and constraints. Consider an analogy. An audition stage is crowded with actors, singers, dancers, and gymnasts, plus a much greater number of people with talents that are irrelevant to the production of a play: football players, chess players, math stars, video-game players, and various talented troublemakers and pranksters of all sorts. In our terms, the "disordered" stage is complex. The director selects the best among them for each role, clears everyone else from the stage, and hands the chosen performers a script. The show is "order." Choosing the performers and imposing rules or writing scripts are not easy. In biology, these are jobs for natural selection. But the diversity of talent—the pure complexity— is easy.

An Evolutionary Drive. Certain nineteenth-century paleontologists thought there were pervasive forces acting within all lineages tending to drive evolution in a common direction: toward larger body size, more ornate and hypertrophied structure, and even greater complexity, in some sense (e.g., Cope 1871). The process was called orthogenesis, and the thinking was that these forces would be enough to overpower selection in some cases and produce poorly adapted organisms—like the famed Irish elk, with its hypertrophied antlers—with a high probability of extinction

(Gould 1977). More broadly, the idea was that all species senesced as they evolved, that they grew old and died, much as organisms do. The twentieth-century modern synthesis vehemently rejected orthogenesis in favor of a Darwinian view that virtually all evolutionary change is adaptation to local circumstances, with no opposition from internal forces. Is the ZFEL an attempt to revive orthogenesis? The answer is both yes and no. It is no in that there is no reason to think that the ZFEL tendency will generally overpower selection, producing poorly adapted organisms, or even that it will ever do so (although in principle it could, depending on the magnitude of the ZFEL tendency and the strength of selection). Also unlike orthogenesis, the ZFEL does not predict any specific changes like increased body size. Rather, it predicts a higher-order sort of change, an increase in variance, which in principle could be manifest morphologically in different ways in every single lineage.

But the answer is also yes. Like orthogenesis, the ZFEL identifies a tendency that acts independently of selection and potentially in opposition to it. Also, the ZFEL is an internal tendency, in the sense that it is the result of the internal redundancy of biological systems and their consequent tendency to differentiate when variation arises. Or, if not internal, the ZFEL is at least not external, not rooted in ecology, not rooted in the relationship between organism and environment, as selection is. Finally, both orthogenesis and the ZFEL tendency are understood to act pervasively. And in the technical language that has grown up in macroevolutionary studies, a pervasive tendency in some group for change to occur in a particular direction is properly called a "drive" (Gould 2002). The drive notion was mostly unwelcome in twentieth-century evolutionary thought on account of its historical association with orthogenesis. The ZFEL rehabilitates the notion, showing how an internal—or at least nonexternal—evolutionary drive is permitted by mainstream theory. Indeed, this is a drive that is *required* by mainstream theory.

General Implications

> Evolution is a change from a no-howish untalkaboutable all-alikeness to a somehowish and in general talkaboutable not-all-alikeness by continuous stick-togetherations and somethingelsifications
>
> **William James (1880), parodying Herbert Spencer's view of general evolution**

The ZFEL is about somethingelsification in biology.[9] It says that somethingelsification is an inevitable background feature of evolution, expressing itself both as somethingelsification among individuals (diversity) and somethingelsification within individuals (complexity). Beyond that, what the ZFEL offers is a unification, a gestalt shift for evolutionary theory, a biological law, and the prospect of a universal biology.

A Unification. The ZFEL reveals a common process underlying diversity and complexity: a common process in the diversification of alleles at a locus in a population, in the differentiation of populations to form subspecies and species, in the diversification and increase in disparity among species and higher taxa, in the differentiation of genes within a genome, in the evolutionary differentiation of cell types, tissue types, and organ types within organisms, and so on. Apparently, disparate mechanisms for these phenomena can be found in population genetics, macroevolution, molecular biology, and morphological evolution. The ZFEL ties them together.

A Gestalt Shift. The ZFEL invites a gestalt shift, a change in what is considered foreground and what is considered background in evolutionary thinking about diversity and complexity. In the conventional view, the background is static. In the ZFEL view, the background is in motion. Then, with a moving background, stasis and decrease shift to the foreground. In the ZFEL view, stasis and decrease in diversity and complexity demand special explanation when they occur; both are revealed to be improbable without the intervention of constraints or forces. The situation is analogous to Newtonian mechanics, in which foreground forces are invoked to explain deviations from background inertial motion. The difference of course is that in Newtonian mechanics, the background state is absence of change—constant velocity—and the deviations are changes in velocity, whereas in biology, for diversity and complexity, the background state is change.

A Biological Law. Biology has been said not to have any true laws, at least not in the same sense that the word is used in the physical sciences. Or if biology has any laws, then it has only one, the principle of natural selection (Brandon and Rosenberg 2003). Obviously, there are empirical generalizations in biology, like Cope's rule that body size increases, on average, or Bergmann's rule, that endothermic animals in colder climates tend to be larger. But one desideratum for laws is universality, and it may be that none of these is truly universal. They may not be true of life on other planets, for example. Different initial conditions could produce very different sorts of evolutionary and ecological patterns. Other generalizations in biology might seem more robust, like Mendel's law of independent assortment, which says that the genes for each character are transmitted independently to the next generation. But these too are probably not universal. Life elsewhere may have utterly different hereditary mechanisms. Indeed, because of linkage of genes on chromosomes, Mendel's law is not even completely true here on Earth: it is more of a rough generalization than a law. Other laws seem to escape this problem, like the Hardy-Weinberg law, which says that, in a large randomly mating population and in the absence of mutation, immigration, emigration, nonrandom mating, and natural selection, gene frequencies and the distribution of genotypes remain constant from generation to generation. But among the many problems that trouble this law, it turns out to be true analytically (to consist of statements of pure mathematics and logic) but to have no empirical content. In contrast, the ZFEL is not true simply as a matter of mathematics or logic.[10] It makes an empirical claim: that diversity and complexity will increase in the absence of constraints or forces, a claim that is testable, not just in principle but in fact. And if the ZFEL is also truly universal, as we argue in the next subsection, it becomes an excellent candidate for a proper law.

A Universal Biology. Certain principles seem sufficiently general and essential that they would have to operate in evolution in any context, for example, in any hypothetical rerun of the history of life on Earth or in the evolution of life on other planets. One commonly acknowledged universal principle is natural selection. In packing a conceptual tool kit to take along to other worlds thought to harbor life, the recommendation is: don't leave home without it. We propose the ZFEL as another. Importantly, the claim is not that the default tendency, the zero-force increase, will be observable always and everywhere. Selection, for example, could overwhelm it. Rather, it is that this tendency will be present and active, observable or not.

There is some reason for confidence that the ZFEL will be true universally, namely that the minimum requirements for it are actually a subset of those for natural selection. It is widely appreciated that Darwin's three conditions—variation, heredity, and differential reproduction—provide the necessary (but not sufficient) conditions for evolution by natural selection. Darwin's achievement was to show that these three conditions are highly expectable, if not universal, and so that evolution by natural selection is an expectable part of life, whenever and wherever it occurs. The ZFEL requires only two of these three conditions, variation and heredity. It applies whenever and wherever these two conditions hold and therefore also whenever and wherever the principle of natural selection applies.

Some might argue that there is so much we do not know about Earthly biology that it is premature, if not presumptuous, to attempt to develop a universal biological theory. Or, in a slight variant of that attitude, some might claim to be interested only in life on Earth. Earthly biology is biology enough. To such attitudes we would counter that universal biology is better suited to explain life on Earth than is a much more narrowly tailored biology. If we start with laws like Hardy-Weinberg, generalizations that are predicated on certain contingent facts about (some of) life on Earth, then we will, at best, be unable to explain the evolution of the conditions (e.g., diploidy, sex) that make such a "law" applicable. At worst, we will be blinded to the very contingency of these conditions. Selection theory has been greatly advanced by its generalization. Our attempt here is to generalize and systematize the many heretofore-isolated consequences of stochastic sampling at every level of biological organization.

New Research Directions

What sort of empirical research might follow from the ZFEL?

1. Quantification of the ZFEL might open the door to new discoveries. In principle, the decomposition of instances of evolutionary change into a vector arising from the ZFEL and another arising from selection (and/or other evolutionary forces) is certainly possible. Indeed, it has already been done at the molecular level (see chapter 3). Consider the simplest case: a lineage of asexual organisms. Now suppose there is a genomic region that is known to be free from selection. Further suppose there are two such lineages that initially have identical sequences in that region. These lineages will diverge from each other at a rate that is easy to calculate. To do this one needs a further empirical fact, namely the mutation rate at each sequence position, which for the sake of this example we will assume is a uniform 0.1 per site per generation. That is it for the empirical side. To calculate a ZFEL-driven rate of change, all we need now is the probability calculus. Thus, the half-life for any site in this situation would be 5 generations. For a bit of genome of length l, approximately 0.5l of the sites should change in 5 generations, 0.75l in 10 generations, 0.875l in 15 generations, and so on.

This example is generalizable. The above calculation of the ZFEL-based rate of change has three components. One is formal, and based purely on the probability calculus, and the other two are empirical. First, we have the empirical fact that the genomic region we are considering is free from selection and constraints. This is the condition for the application of the special formulation of the ZFEL. And there is nothing a priori about whether or not this condition is met. Second, there is an intrinsic rate of change for the entities in question—in this case nucleotides. This is clearly an empirical parameter.

Can we apply the same methodology at a higher level, to the phenotype? Consider the sort of species diversification in morphospace discussed in chapter 3. To keep our example analogous to the one above, let us suppose that we have two sister species that are initially morphologically identical. This supposition is not as clear as it might sound. For genome sequences, we have a universal space for comparing sequences, based on the shared nucleotide alphabet. But at present we have no universal morphospace into which we could place two sister species. As discussed in chapter 3, there is no reason in principle why such a space could not be developed. However, for present purposes, we really do not need a universal morphospace. For two sister species, a more restricted morphospace would suffice. If the species were mammals, then we would only need a morphospace defined by some set of shared mammalian characters.[11] In effect, our project would be to quantify the effect of the ZFEL insofar as it acts on just that set of characters.

Now, by supposition, our two species start at an identical location in morphospace. How will they diverge over time, based on the ZFEL? Following the above molecular example we could construct a quantitative model of ZFEL-based change.[12] There would be a formal part based on probability theory. Then we would need two empirical parts. First, we would need the empirical data that support the application of the special formulation of the ZFEL to this situation. Basically, we would need to know that some sets of characters were evolving randomly with respect to each other, at least to some extent (see discussion in chapter 3). We would then need to empirically measure some intrinsic rate of phenotypic change in that set of characters. Given these two empirical inputs, a quantitative model of ZFEL-based change could be produced.

We do not pretend that any of this is easy. We simply want to make the point that quantification is doable in principle. Further, we could expect significant payoffs, of the same sort that have come from this sort of work in molecular evolution. That is, we could then identify the forces that are either working in concert with the ZFEL or opposing it. If novel macroevolutionary forces exist, this would be the way to identify them.

2. The ZFEL offers a basis for predicting where and when increases in diversity and complexity are likely to occur. It tells us to look for increased diversity in taxa and ecological circumstances where selection and constraints are likely to be reduced and to look for increased complexity in organismal features that have been removed from selection for functionality. To some extent, these are predictions of things we already know. Diversity increases after mass extinctions, when ecological constraints are reduced. Genome complexity increases when genes produce functionless duplicates. But the ZFEL also makes novel predictions. It tells us to expect diversity to increase even when no special ecological opportunities can be identified, as in geographically isolated populations, and even when no environmental variation is present, as in isolated species living on a flat, apparently unvariegated sea floor. And it tells us to look for complexity to increase even when there is no clear advantage to differentiation of parts, as in the eyes of cave fish.

We leave the invention of other predictions to the imagination of the reader and instead turn to diversity and complexity in human institutions, organizations, and culture generally. The human case is not the focus of our work, so we make these remarks only in passing, in a speculative frame of mind, and as a spur to debate. Our species is said, sometimes, to be moving toward homogeneity, culturally as well as genetically, especially in recent decades. The existence of forces producing homogeneity is undeniable. However, the ZFEL predicts that wherever these homogenizing factors are absent or reduced, diversity will spontaneously arise. Given appropriate measures of the flow of cultural elements (ideas, languages, technologies, etc.), as well as measures of cultural diversity, this seem eminently testable. The ZFEL also tells us to look for complexity to increase, even when there are no gains in efficacy or efficiency and even when there are no profits to be had. In the absence of counterforces, organizations of all sorts—from large nations and businesses to small communities and clubs—are expected to become increasingly complex (in the pure sense) over time. Unless constrained, they are expected to acquire ever more "part" types, that is, to consist of subunits that are ever more differentiated from each other. This too seems testable.[13]

3. Finally, there is the question of relative frequency. How often is the ZFEL an important factor in the evolution of diversity or complexity? Is it always something we need to take into account? Or perhaps it is usually overwhelmed and rendered negligible by other factors. At issue is the importance of the ZFEL in evolution. Here we address two concerns that a detractor might raise, two arguments that the role of the ZFEL in

evolution has been quite limited, the second suggesting a possibly pro-
ductive line of research.

A critic might say that the conditions for ZFEL-driven complexity
and diversity should be limited to cases where selection is reduced or
absent, such as those discussed in this chapter, and that these are expected
to be rare. But this charge misunderstands the ZFEL claim. The cases
we have been discussing are just those in which the ZFEL is most easily
observed. The general statement of the ZFEL says that a tendency for
complexity and diversity to increase is always present, in evolution generally. Consider the relationship between moving objects and Newton's
law. An inert object moving alone in intergalactic space feels very little
gravitational pull and is ideally positioned to reveal the effect of inertia uncomplicated by outside forces. But that does not mean inertia is
unimportant for other objects, including those buried deep in solar systems
and buffeted by multiple powerful gravitational fields. Inertia is important, a critical part of the calculation of the trajectories of all objects,
even if it turns out that the vast majority of all objects are powerfully influenced by gravity, even if the zero-force condition occurs very rarely.

Alternatively, a critic might argue that even if the ZFEL acts pervasively, its effect might nevertheless be small in most contexts when
compared with the effect of selection and constraints. This is certainly a
possibility in principle, although it is worth pointing out that the ZFEL
acted quite powerfully in the two cases discussed in chapters 3 and 4,
macroevolutionary diversity and pseudogenes, and further that the
domains of these two cases are quite large. Even if the ZFEL were important nowhere else, those two together make it fairly significant in evolution. Still, the objection is a fair one. To properly answer the question
of relative frequency, ideally one would have to be able to decompose
a large number of instances of evolutionary change in diversity or
complexity, or failures to change, into their ZFEL components and those
due to other causes. Ideally, these components, including the ZFEL, would
be quantifiable. Then, to assess relative frequency, one might measure
the contribution of the ZFEL in some arbitrary sample of instances, a
sample covering a range of organisms, taxonomic and temporal scales,
and ecological circumstances.[14]

What Explains Diversity and Complexity?

Let us reconsider the question raised in chapter 1: what explains diversity
and complexity? Diversity is widely recognized to be at an all-time high,
at least for animals, and impressionistically at least, the modern biota

contains the most complex organisms of all time. Both are marvels, of a sort, and we naturally seek an explanation for the trend they imply over the history of life. The ZFEL by itself does not answer this question. But it does tell us what we need to find out: the magnitude of the ZFEL tendency and the magnitude and direction of selection, as well as those of any constraints. At present, all are unknown. Thus, our central claim here is not that the ZFEL has driven the rise in diversity and complexity over the history of life, although it could have. Nor is it even that the ZFEL has been an important factor in the evolution of diversity or complexity, although it probably has been. Rather, it is the more modest-sounding assertion that standard selectionist explanations are in principle incomplete. We say "modest-sounding" because we mean *all* selectionist explanations for diversity and complexity. The scope we claim for the ZFEL is immodestly large. The claim is that the ZFEL tendency is and has been present in the background, pushing diversity and complexity upward, in all populations, in all taxa, in all organisms, on all timescales, over the entire history of life, here on Earth and everywhere.

Notes

1. One of us cannot resist the urge to point out—and the other does not dispute—that the Red Sox managed to end that streak only by adopting, as James Fenimore Cooper (nearly) put it in *The Last of the Mohicans,* "the ways of the Yankées." Still, the other points out, they won, and won gloriously, and that's that.

1. Here and throughout, we use the word "variance" in a generic sense, to refer to something like "amount of variation" or "degree of differentiation," *not* in its statistical sense, a sum of squared deviations from a mean. On the other hand, as will be seen, the statistical concept is often a good measure of variance in the generic sense. We discuss this issue more later.

2. For example, a bias in the direction of mutation might be considered a force.

3. Throughout this book, we use the acronym ZFEL (pronounced "zeff-el") to refer generically to both formulations or, where it is clear from the context which we intend, to refer to just one. Where the referent could be unclear, we identify the first as the special formulation and the second as the general formulation. Whenever "ZFEL tendency" is mentioned, the reference is to the general formulation.

4. The ZFEL draws on the same underlying principle that motivates Lynch's work and more generally can be seen as part of the recent turn that molecular biology has taken in acknowledging the importance of random processes. Indeed, on account

of this shift, we expect the ZFEL to find its most sympathetic audience among molecular biologists. Nevertheless, for clarity we need to point out that the focus of this turn has always been on adaptation, more precisely on the importance of random processes in the production of adaptation and adaptive novelty. As will be clear shortly, our interest is different.

5. For empirical surveys of diversity, see Knoll 2003; Alroy et al. 2001. For complexity, see McShea 1996; Carroll 2001; Sealfon 2008.

6. This is not to say that anything in complexity theory is inconsistent with our approach, just that the research program is different.

7. It is tempting to use the phrase "levels of organization" here, but the word "organization" would be misleading in that our understanding of hierarchy does not require that the parts of an entity at a given level be especially organized, at least not in any functional sense. For example, in our usage of the word "hierarchy," a clade occupies a hierarchical level above the species that constitute it, even if those species are not especially integrated, or organized. Our notion of hierarchy is close to what Salthe (1985) called a "scalar hierarchy."

8. We do not mean to imply with this list that we recognize only a single hierarchy in biology. Multiple and somewhat-overlapping hierarchies have been recognized (e.g., Eldredge and Salthe 1984) in biology. In our understanding, the ZFEL applies at all levels, in all hierarchies of objects, so long as they exhibit heritable variation. So, to pick a potentially problematic case, one might ask whether the ZFEL applies to ecosystems. The answer depends on whether ecosystems reproduce and, if so, whether they do so with any fidelity, so that variation among them would be to some degree heritable. To the extent that they do, the ZFEL applies.

9. Probably this usage of "complex" will sound less odd to biologists who study social insects or other highly social animals. For them, the idea that a complex group is a group with a diversity of types of individuals should be familiar.

CHAPTER TWO

1. One of us experimented with introducing various patterns of correlated variation into a series of measurements taken from serial structures, simulating the effect of the introduction in evolution of various gradients and fields (McShea 1992, 2005b). For example, a sine-wave function with a fixed period and amplitude was added to a series of measurements of lengths of vertebral bodies in a single mammalian vertebral column. Interestingly, standard deviation, range of variation, and other measures that could be interpreted as measures of pure complexity increased on average, and they did so with the addition of any of a variety of simple functions and over a wide range of parameters attached to those functions. This is not to say that functions producing decreases cannot be engineered. They can. The point is that most arbitrarily chosen functions seemed to have the opposite effect. This finding suggests a general principle, perhaps worth a more serious mathematical investigation than we are competent to give it: given some set of measurements and an unbounded measure of the diversity among those measurements, most arbitrarily chosen functions (representing patterns of correlated change) that can be imagined will have the effect of increasing diversity by that measure.

2. As we discuss in chapter 3, the expression and detection of the ZFEL can be confounded not only by forces and constraints acting in opposition but by those that act in the same direction as well. Correlations can *augment* the ZFEL effect, as well as oppose it. In the human dispersal example, the variance in their locations would increase rapidly if the people were actively fleeing each other for some reason. In biology, selection on two species for avoidance of competition might produce differentiation. In neither case is the resulting dispersal or differentiation the result of independent variation, of entities moving or changing randomly with respect to each other. Or at least, it is not solely the result of this, and to the extent that it is not, the dispersal and differentiation cannot be properly attributed to the ZFEL.

CHAPTER THREE

1. As an aspect of variance, diversity is a concept that applies only to entities with at least two components, for example, populations with at least two individuals. For populations consisting of only a single individual, diversity is zero by definition and therefore not informative.

2. Indeed, paleobiology has often used taxic measures—counts of taxa—explicitly as mere proxies for disparity in cases where the continuous morphometric measures are difficult to apply (Foote 1997; Erwin 2007).

3. A seeming objection is that certain measures can be imagined under which the ZFEL prediction seems to fail. For example, consider a phenotypic variable that is expressed as a percentage, say the percent coverage of some surface by a pigment. And suppose that the population is initially distributed bimodally, with half at each extreme (i.e., half with almost no pigment and half with 100% coverage). Now, if we adopt some discrete measure of evenness as our diversity measure, then diversity will be measured as low initially, with all individuals concentrated at the ends of the percentage range, and diversity will increase, as predicted by the ZFEL, as the population spreads from both ends toward the middle. But if we adopt the statistical variance as our measure of diversity, then diversity is maximal initially and declines with time, as the population spreads, apparently contradicting the ZFEL. However, notice that this result is a consequence of the starting position adjacent to two mathematical limits, 0% and 100%, limits that happen not to affect a measure of evenness (at least initially, although later they will cause evenness to plateau) but that affect the statistical variance instantly. Here we take these limits to be a kind of constraint, one that is built into the structure of the space in which we have chosen to measure diversity. And on account of this constraint, the special formulation of the ZFEL does not apply. Applying the general formulation, one would say that in a constrained space, an increasing *tendency* is present but may be unrealized, owing to the constraints. Notice that, if we had used the actual area of coverage (or the log of area covered, to eliminate the boundary at zero) rather than percent coverage, constraints would be absent, and diversity would have increased by both measures, and further, it would do so indefinitely. We raise this issue again later in this chapter in our discussion of genetic drift.

4. In privileging disparity, we do not mean to imply that we think the discreteness of taxa is somehow biologically unimportant. On the contrary, one of the most interesting puzzles in evolutionary theory since Darwin has been the origin of the discontinuities that produce discreteness. It is simply that the ZFEL is interested in something else, in the production and accumulation of variation, rather than the causes of discreteness, and it turns out that continuous measures are applicable (somewhat) more generally for this purpose.

5. The picket fence in chapter 1 is an imperfect analogy in that both reproduction and heredity are absent. Still, picket fences do accumulate variation and therefore show a ZFEL-like increase in diversity. And so it is not hard to imagine a ZFEL-like principle that applies to nonliving systems in which the requirement would be for analogs of reproduction, something like "persistence," and of heredity, perhaps "memory." A picket fence has persistence and memory and therefore accumulates variation in a way that, say, a puddle of water does not. See chapter 6.

6. Notice that the "death and extinction" constraint is a special case of the population-size constraint mentioned earlier in the discussion of discrete and continuous senses of diversity. When population size is static, the ZFEL predicts increase in the continuous sense and also in the discrete sense (but only until every individual is a unique taxon). When the population is decreasing, diversity is predicted to decrease, ultimately, by both measures. And when it is increasing, the population-size constraint is removed, or at least reduced, and the special formulation says that diversity will rise in both senses.

7. When talking about phenotypic selection, the term "stabilizing selection" is used, while molecular evolutionists use the term "purifying selection" to describe the sort of selection that eliminates most mutational variants of a given gene (as in the *Pax6* case). Here we treat them as synonyms.

8. At the phenotypic level there has been a major metastudy of forms of selection in natural populations (Kingsolver et al. 2001). Surprisingly, that study did not show a high prevalence of stabilizing selection. But the authors suggested, and we agree, that this is most likely an artifact of a weakness in the sorts of studies surveyed rather than a reflection of how selection operates in nature.

9. Admittedly, the focus of these analyses has been, not on divergence, but rather on homoplasies (shared similarities not due to common descent), which presumably reveal the action of selection. But this methodology makes sense only if *divergence,* or *failure to converge,* is the background expectation for lineages *in the absence of selection* for similarity, in other words, the ZFEL.

10. In the experiments that have been done with null models of disparity change (e.g., Foote 1996; Gavrilets 1999; Pie and Weitz 2005), the focus has been on explaining the apparent deceleration in the rate of morphospace occupation over the Phanerozoic. Our point here is that disparity increases monotonically when constraints and selection are absent, a claim that all of these models support. Now, in some versions of these models, such as those in which diversification is accompanied by very little or no morphological change, or in which morphological change is otherwise limited, disparity can asymptote, or actually decrease, by certain measures. A relatively small morphospace can become ever more densely populated with lineages, producing a decrease in, say, the average

difference among pairs of lineages. However, these cases are not exceptions to the ZFEL, in that constraints—limits on the amount of morphological change allowed in each lineage in each time step—are present. Extinction too can cause disparity to decrease in these models. But extinction is also a constraint (as discussed earlier in this chapter) and, therefore, does not contradict the ZFEL either.

11. The details of Sepkoski's findings have been under discussion lately, especially the magnitude of the increase in the number of marine taxa during the Mesozoic and Cenozoic Eras, the last half of the history of animal life (Peters and Foote 2001; Alroy et al. 2001, 2008).

12. This finding agrees with and extends Bambach's (1983, 1985, 1993) earlier finding that the number of guilds in marine animals increased from the Paleozoic to the Cenozoic.

13. Despite the evidence, it is possible to question the increase in metazoan macroevolutionary disparity. For one thing, there is some evidence that most of the increase may have occurred in a short burst at the start of the Phanerozoic, with a significant decline in rate occurring after that. For example, Thomas, Shearman, and Stewart (2000) found that a large proportion of a space defined by the form of animal hard parts, the metazoan "skeleton space," filled up early in the history of animal life. (See also Valentine 1969 and Foote 1996.) Also, Gould (1989) famously argued that arthropod disparity was high early in metazoan history (notably among the arthropods of the Burgess Shale, about 500 million years ago) and that modern arthropod disparity is relatively low. In particular, the Burgess contains a number of oddballs with morphologies well beyond the limits of the modern arthropod classes, while the modern species are all very similar to each other, on average. He went on to argue that the same basic pattern of change—high early disparity with either little increase or decline later—characterizes animal evolution generally and, further, that this pattern suggests a predominance of developmental constraints over natural selection in post-Cambrian evolution. Gould's claims have generated a modest debate, in which both the pattern and his explanation for it have been challenged (Briggs, Fortey, and Wills 1992; Conway Morris 1998, 2003; Briggs and Fortey 2005). We take no sides in this debate. But we can say this: if disparity increases but the rate at which it does so declines, the decline in rate requires an explanation but does not contradict our premise here that disparity increases. On the other hand, if disparity actually decreases, then our argument in this section—that increasing disparity is evidence for the ZFEL—dissolves. Still, even in that case, our larger point with regard to disparity is that, in the absence of opposition from selection or constraint, disparity is expected to increase with the passage of time. And that increase is the null expectation (Foote 1996; Gavrilets 1999; Pie and Weitz 2005; Erwin 2007). It is the ZFEL, and all parties to the debate seem to accept it. Indeed, to our knowledge, in the history of the debate, that point has never been challenged.

14. Independence among lineages is imperfect, of course. In particular, similar environmental pressures and opportunities may lead to similar sorts of evolutionary changes. But no two lineages are expected to respond in exactly the same way, because each has a unique phenotype and evolutionary potential, and therefore each experiences and responds to these similar pressures and opportunities to some degree differently. To the extent that this is the case,

divergence occurs. Some might argue that the history of life is dominated by massive convergence, or homoplasy, and parallelism at all taxonomic and temporal scales, driven mainly by selection (e.g., Conway Morris 2003). But true or not, the point is not relevant here. It would be relevant if our goal were to *predict* whether the ZFEL applies to the history of metazoans and whether disparity is therefore expected to increase. To the extent that selection (or developmental constraint) produces homoplasy and parallelism, changes are not independent among lineages, and the special formulation of the ZFEL does not apply. However, our point is different. It takes for granted that standard intuitions, as well as the technical findings of Novack-Gottshall and Bambach et al., are correct, in other words, that disparity has increased. It takes for granted that independence among lineages, however imperfect, has been sufficient to allow them to diverge, on average. And the point is to explain the observed divergence.

15. Interestingly, the ZFEL *is* involved in many of the standard microevolutionary mechanisms for divergence. The mechanisms of speciation that Schluter (2009) groups under the heading of ecological speciation implicitly invoke the ZFEL. For example, speciation occurs when two populations of a species are subject to different selection pressures and diverge as a result, both genetically and phenotypically, leading to reproductive isolation. Speciation by this route has been demonstrated, both in the lab and in nature, and is thought to be common in nature (Schluter 2009). It is also the ZFEL. Both populations are under selection. Selection is the cause of change in both. But differentiation is the incidental result of differing selection pressures. One of the standard explanations for divergence in Darwin's finches of the Galapagos Islands invokes this mechanism. Speciation can also occur by what Schluter (2009) calls mutation-order selection, in which different mutations arise—or arise in a different order—and become fixed in different populations of a species, even though both are subject to similar selection pressures. There is evidence for this set of mechanisms as well, although its frequency in nature is not known. This too is the ZFEL, in that change in each population occurs by mechanisms that are independent of those in the other. And again, reproductive isolation and speciation—and therefore diversity—are the result. We will not venture a view on whether these ZFEL-driven mechanisms predominate in microevolution. But we will note that—unlike microevolutionary selection for divergence—propagation to higher levels is unproblematic.

16. What would be required for selection for divergence at the population or species level to propagate to the highest level? It would have to be that selection favored differentiation among all or most species in some shared character or set of shared characters. For example, suppose every species on the planet were under selection to have a unique body size (in other words, to be as different as possible in body size from every other species). In that case, the resulting disparity among all species along a body size axis would not be the result of the ZFEL. On the other hand, disparity arising from differentiation in nonshared characters— which in higher taxa would inevitably be *most* characters—would still be attributable to the ZFEL.

17. We note that the ZFEL is immune to this critique. The ZFEL does not require ecological interaction of any sort. All it requires is variation and heredity, and these two qualities are abundant at the species level and higher.

18. For some traits, such as versatility or evolvability, the consequence seems to be both reduced extinction probability and increased origination probability, and for these, selection clearly favors diversity. For others, however, such as population density or abundance, extinction and origination probabilities seem to change in the same way, so that the consequences of species-level selection for diversity are ambiguous (see table 1 in Jablonski 2008).

19. Jablonski's data might represent an interesting exception to the general rule. During mass extinctions, the rules of the game might change dramatically so that clades, which are essentially genealogical entities, can become ecological actors.

20. For example, suppose we measured disparity as average distance from the mean in an appropriate morphospace. In that case, the disparity among species within each class, measured as average distance of the orders in that class from the class mean, might be very high, while the disparity among classes, measured as the average distance of the class means from the phylum mean, might be very low. This would be the case if the class means were highly clustered.

CHAPTER FOUR

1. Interesting relationships have been proposed between absolute numbers of parts and differentiation, notably Williston's law, which proposes that absolute numbers decrease as differentiation rises (Williston 1914; see also Buchholtz and Wolkovich 2005).

2. Or precisely 92, if we include elements present in trace amounts and limit ourselves to the naturally occurring ones.

3. The conceptual scheme underlying pure complexity is actually more complicated than we have let on. See the conceptual scheme laid out in McShea 1996. In that scheme, what we here call pure complexity refers not just to part types but to complexity in any function-free sense. In addition to the horizontal and vertical aspects, there is also the pure complexity of processes—for example, the number of steps in the generation of an object, which for an organism is its development and for a machine its manufacture. There is the pure complexity of the spatial arrangement of parts, irregular being more complex than regular. And there is also the pure complexity of their pattern of interaction. And so on. Three points: (1) All are "pure" in the sense that they are independent of function. (2) All are conceptually independent of each other. A thing can be vertically complex and yet be horizontally simple (e.g., Chinese boxes, if there is only one part type at each level). A thing can be horizontally simple but generatively complex (e.g., mayonnaise is a simple homogeneous gel at normal scales of observations but some recipes for making it are enormously complex). And so on. (3) All may be of interest in some context, but here we are interested only in horizontal complexity and only the horizontal complexity of parts. Processes are not considered. Patterns of interaction are not considered. Spatial arrangement is not considered. We can imagine that a somewhat-different version of the ZFEL might be devised that is applicable to some of these at least—a principle that predicts that processes should become more differentiated, patterns should become more irregular, and so on. But here we are concerned only with parts.

4. If these examples seem problematic, we suggest two sources of misunderstanding. (1) The first is scale slippage. It might seem that before we can declare roses and chrysanthemums more different from each other than roses of two different colors, we need to know something about the fine-scale structure of these flowers, perhaps at the level of genotype. But recall that pure complexity is a level-relative concept. Thus, the claim is only that the two flower species are more different at ordinary scales of observation, say, for human beings examining them by eye from a distance of a few feet. There is no implied claim that roses and chrysanthemums are more different in their small-scale anatomy or even in their genes (although they probably are), nor would the discovery that they are not contradict the larger-scale observation that they are. (2) The second potential source of misunderstanding is function slippage. In the knife example, we seem to be treating each type of cutting tool as a single unit, comparable as a whole to other such units, and ignoring possible complexity differences in the composition of each. A single blade, for example, could be quite complex, consisting of an integrated set of functional units, such as a point for piercing, an edge for cutting, a spine for stiffening the point and edge, and so on. This concern is misplaced in two ways. First, it involves scale slippage. The comparison is at the scale of blades as wholes, not at the scale of their components. Second, pure complexity is deliberately indifferent to function at all scales. A single blade is a single piece of metal and therefore counts as a single part, at ordinary scales of observation. If we were measuring complexity at a smaller scale, we might want to decompose it into subparts—point, edge, spine—based on variation in composition and shape but not based on function. Pure complexity is function free. To dramatize this difference, consider that, in our view of complexity, defects in the blade—such as regions of greater wear that reliably form on the blade—are functionless but would count as differentiation and would therefore contribute to complexity at the small scale.

5. For example, Crutchfield and Young (1989) and Crutchfield (1992) have argued that systems contain both a regular and a random component and have suggested that complexity is the degree of differentiation in just the regular portion. In this view, random refers to unique features, such as the precise number of hairs on the arm of a single human individual, while regular refers to shared features, such as a five-fingered hand. The intent is to restrict complexity to features that are "rule based," in other words, to those produced either by natural law acting in the present or by irrevocable, contingent events in the past (frozen accidents) (Gell-Mann 1994). This understanding of complexity might sound very different from ours but it is actually fully consistent. A decision that two cells are the same type can be construed as a decision that their similarities are rule based and their differences are not. More generally, to identify types of parts is to discern first-order regularities.

6. Technically, complexity is a concept that can be usefully applied only at a level smaller than the whole. Thus, an organism consisting of a single part—say, a single-celled protist, if we are measuring complexity at the cell level—might be said to have a complexity of 0 (with complexity understood as degree of differentiation) or 1 (complexity as part types). But this would be true by definition, and therefore the number conveys no useful information.

7. In our own thinking about complexity, we have found it easy to slip unintentionally from a complexity discourse into a diversity discourse. Thus, it is worth restating that the issue for complexity is variance, not among individuals, but among parts within an individual. So for complexity, the prediction of the ZFEL is not that the parts of individual A will be more different from the parts of individual B than the same parts were in their parents (although that *is* a prediction for diversity). Rather it is that, if individual B is descended from individual A, the parts of individual B will be more different from each other than the parts of individual A were from each other.

8. Here, and in the particle model in chapter 2, the assumption is that parts vary independently and further that variation is introduced independently to each element. This assumption, however, is not realistic for organismal development, where for many parts, such as iterated parts, covariation is the norm. Indeed, for some iterated structures, the developmental mechanism underlying covariation is reasonably well understood (e.g., Pourquie 2003), or at least well enough understood to support the suggestion that many perturbations of development will not produce completely independent variation among the parts. Two things need to be said here. First, it may be that most arbitrary perturbations of covariance functions will tend to produce an increase in differentiation nevertheless. As discussed earlier, one of us (McShea 2005b) examined the effect of introducing various patterns of covariation to a series of measurements of parts, and in most cases, complexity increased. Second, and whether or not the latter is general, *complete* independence is unnecessary. ZFEL-driven differentiation is expected *to whatever extent parts develop and are perturbable independently,* in other words, as long as the degree of independence is nonzero. And we can say with confidence, based on first principles, that independence will be nonzero simply because each iterated part differs in spatial location from every other and therefore finds itself to some extent in a unique developmental environment, subject to unique perturbing factors, at least some of them inevitably heritable (Spencer 1900; West-Eberhard 2003).

9. A required assumption here is that there is no selection for differentiation itself. That is, we assume that each claw is selected independently for its own special function, and that the differentiation is not driven by the advantages of "differentness," so to speak.

10. Parts, as understood here, are different from what have lately been called "modules" in biology (e.g., Wagner and Altenberg 1996; Callebaut and Rasskin-Gutman 2005). The focus of the modularity literature has been mainly on the emergence in evolution of patterns of covariation in *development,* that is, evo-devo modules. Parts, in contrast, are patterns of interaction in an organism on much shorter timescales, what might be called mechanical, physiological, or behavioral "modules." There are theoretical reasons to think that parts might line up well with evo-devo modules, but this has yet to be established in organisms. Here what matters is that they are different conceptually (McShea and Venit 2001; McShea 2000).

11. Notice that the claim that parts evolved as functional modules does not even make sense without admitting and covertly employing a function-free notion of parts.

12. It might seem that there is a genetic shortcut available here. Why bother to count parts when we can just count genes? The thinking here is that genes construct the parts, and therefore, the complexity of the genome is a fair proxy for parts complexity. The answer is, first, that genes do not construct the parts; rather, they participate in that construction, along with physical laws and innumerable environmental factors. Second, the genetic proxy is hierarchically problematic. What level of organization are genes supposed to be a proxy for? Is number of genes a proxy for number of molecule types, number of cell types, number of tissue types, . . . ? Because complexity is a level-relative concept, so that different complexity values are expected at different levels, there can be no principled answer to this question. And third, the underlying assumption of a genetic proxy may be wrong. We really do not know the nature of the relationship between genes and parts. An increase, say, in genome complexity could be transduced by development into either an increase or a decrease in parts complexity. More gene types could in general be associated with more part types, but in any particular case, or even in *many* particular cases, it could also be associated with fewer. (See next section.) And therefore, if our interest is in the complexity of an organism at the scale of its organs, say, and not in the complexity of the genome itself, there is at present no substitute for counting organs.

13. Obviously, the ZFEL is not expected to operate at levels where the parts themselves are not variable. In an organism, an oxygen atom and a carbon atom are two parts at the atomic level, but those atoms will not tend to become more different from each other, at least on timescales normally experienced by organisms.

14. It is probably worth recalling here that the ZFEL is concerned with complexity of structure, not of function. Thus, the issue here is *not* gain of function versus loss of function, familiar from molecular biology. Rather, it is gain of parts versus loss of parts, with no assumption that parts must be functional.

15. There is an older literature on the tendency for structures to be "reduced" in the absence of selection (e.g., Brace 1963; Prout 1964). This phenomenon is irrelevant to the ZFEL when "reduction" is understood to mean only reduction in size because a structure could become smaller without losing complexity if all of its parts survived. In fact, however, what is often seen with reduction is a loss of parts as well, or of differentiation (Wilkens 2007)—hence the conventional wisdom that random variation reduces complexity. This may be correct, but to our knowledge, no systematic study has been done to show that it is true in general.

16. It is worth mentioning a recent study by Lohaus et al. (2007), which was presented as a test of an earlier version of the ZFEL (McShea 2005a, 2005b) and produced results that appeared to them to contradict it. Lohaus et al. examined a model of a developmental system and found a tendency for complexity to increase in simple versions of their model but a tendency to *decrease* in versions that were already complex. No selection was applied, but this fact alone is insufficient to refute the ZFEL. Their model was a complicated one, potentially rich in constraints that might bias the result against complexity. Notice that we are not claiming that their model is unrealistic. It could be wonderfully realistic,

and to the extent that it is, Lohaus et al. may have gone some way toward demonstrating that, in complex real organisms, development may be biased toward simplicity. But we need to point out that their finding does not contradict the ZFEL, as formulated here. The ZFEL claims that complexity increases when selection and constraints are absent (special formulation). It does not claim that these conditions must be met in real organisms, and therefore, it would be unsurprising if a realistic model failed to reveal the predicted tendency. Indeed, we think a major virtue of the modeling approach is the potential it has, perhaps revealed in the Lohaus et al. study, to suggest candidates for generative biases in real organisms, biases that could oppose the ZFEL but also those that could augment it.

17. It may be helpful to remind the reader here that the ZFEL as we have formulated it is concerned only with objects, and therefore, we are treating genes here as objects. Of course, genes are objects with critical roles in processes of various kinds. And it may be that another version of the ZFEL could be devised that predicts increasing differentiation among processes and interactions in the absence of selection and constraints. There could be a ZFEL for complexity in other pure senses. See above, n. 3. But that would be a different project.

18. It bears restating here that the issue is complexity, not diversity. Increasing complexity of the genome is the differentiation of genes with respect to other genes *within an individual* and should not be confused with the differentiation *among individuals* of genes at the same locus. The ZFEL predicts both, but they are different.

19. The work of Lynch (2007b) and allied workers in molecular biology has been inspirational for us and also is a critical part of the case we are making here for the ZFEL at the molecular level. Their central argument is that neofunctionalization is a two-step process, the first step of which is randomization and the second step of which is a selection-driven acquisition of function. In effect, the first step invokes what we are calling the ZFEL. More generally, they argue that nonadaptive processes have played a central role in evolution. And this too is consistent with the spirit of our proposal and seems to us right on target. (Indeed, our proposal must be—in some difficult-to-trace way—partly derivative of these arguments.) Still, it may be worthwhile to point out some differences. First, the ZFEL is concerned only with diversity and complexity and not with any other properties of the genome. Second, the ZFEL is not (directly) concerned with the second step, acquisition of function. Third, the ZFEL applies at all hierarchical levels, not just at the level of DNA. And fourth, our main mission is to advance a hitherto-unrecognized unifying principle and only secondarily to argue that it has been an important factor in evolution.

CHAPTER FIVE

1. Although philosophers disagree about the value of old evidence (evidence known before the formation of the hypothesis) versus new evidence, the stance we take here is totally uncontroversial. We hold that agreement with data, whether old or new, is a good thing, and that the ability to predict heretofore-unknown phenomena is also a good thing.

2. Notice that we are using nonselected genes to make a different point than in the pseudogene case. Here the point is the increase in differentiation among individuals or taxa, that is, diversity. In contrast, in chapter 4, we used the pseudogene case to show an increase in differentiation among genes within an individual, that is, complexity.

3. See http://jaxmice.jax.org/geneticquality/stability.html. Also see Bailey 1977.

4. It could be that we do not need even reduced selection to see the ZFEL in action. The reason is that at present we have as much reason in theory to think that selection opposes complexity, on average, as to think that it favors complexity. The same is true for the structure of development. Given our ignorance on these matters, a fair starting assumption might be that selection and developmental structure are complexity neutral, on average. And if they are, simple parent-offspring comparisons under natural conditions should reveal the effect of the ZFEL.

5. For the atlas and the axis, a variation would add a novel part type but the gain would be accompanied by the loss of the original type. For example, a variation in the atlas generates a novel part type (an atlas with a new morphology) but also destroys an existing part type (the old atlas morphology), resulting in no net change in complexity. On the other hand, if we did not restrict ourselves to discrete variation, if we measured differentiation on a continuous scale, the ZFEL tendency to differentiate could reemerge, with most variations arising in the atlas, for example, tending to make it more different from other cervicals than it already is.

6. Ehling observed deformities in the control animals as well as the irradiated ones, and one might imagine that radiation-induced variation would differ in some systematic way from natural variation. As it happened, in this case it did not, but even if it had, the source of variation is not important here. The ZFEL point is that morphology is strongly redundant, and that the effect of variation of any sort is to destroy redundancy, producing differentiation. Thus, unlike Ehling, we are indifferent to the source of variation, and therefore we combined experimentals and controls. Four of the 10 animals were controls.

7. The ZFEL prediction is consistent with several alternative hypotheses about the mechanism underlying evolutionary change in the eyes of these cave organisms. An increase in complexity could be the direct result of the accumulation of mutations. Or it could be the incidental result of adaptive changes in other structures with developmental linkages to the eyes. In the latter case, the changes in the visual system arising from life in the dark would be the result not of the absence of selection but of selection that is unrelated to vision and is therefore *random with respect to* the selection pressures operating on normally sighted cave fish. As discussed, the ZFEL treats these as equivalent.

8. The literature on regressive evolution in cave animals is extensive, but the focus appears to have been on divergence *among* individuals and among species, both in morphology and in genes, rather than divergence of parts *within* individuals.

9. Other mechanisms producing trends are known and have been discussed in the literature, some involving variation in the direction and magnitude of the

rightward tendency or variation in speciation and extinction probability, along the horizontal axis (Wagner 1996; Alroy 2000; Gould 2002). For present purposes, restricting the discussion to the three mechanisms discussed in the text—which have been focal in the literature—is invaluably simplifying.

10. Gould's book *Full House* has been misinterpreted as arguing that the trend in complexity was the result of chance, that it was an improbable and unrepeatable aspect of life's history. Gould did argue in an earlier book, *Wonderful Life,* that many features of modern organisms are improbable, that they are frozen accidents, of a sort that could easily not occur in a rerun of the "tape of life." But in *Full House* his point was different. It was about the *mechanism* of the trend in what he called "excellence" and we call colloquial complexity. Gould thought the mechanism was passive, but as he would undoubtedly have acknowledged, whether it was or not, there was nothing improbable about it. The trend itself—properly speaking, an increase in the mean—would be virtually inevitable in reruns of the tape of life under any of the three trend mechanisms.

11. We point out, however, that the assumption that modern and ancient bacteria are both simple and about equally so, in the colloquial sense, has not been tested. One reason is that colloquial complexity is not measurable, of course. Another is that very little is preserved of the first ancient bacteria beyond their gross morphology.

12. In making this point, we put aside the possibility that viruses and prions might be considered alive, and simpler than bacteria. We also put aside the possibility that there could be yet simpler living entities, still undiscovered, either in the past, the present, or both. If either of these possibilities were taken seriously, one would have to admit the possibility that the minimum has increased (or even that it has decreased). Here we simply adopt the conventional view that it has not.

13. Gould (1996) deserves considerable credit for explaining the passive mechanism so artfully and raising it to the level of plausibility. And his two main observations, the persistence of bacteria and their current diversity (the idea that bacteria have long occupied the statistical mode in the history of life), do seem to rule out a strongly driven mechanism. But he gives insufficient weight to the possibility of a weakly driven one, which is not ruled out by the persistence of simple organisms.

14. In discussions of the driven mechanism, McShea (1993, 1996) invoked selection as the likely cause underlying strongly driven trends, wherever they occur. This may be a fair inference for trends in most features of life, but in light of the ZFEL, it is clearly not for complexity. It is evident now that a strong drive could just as well be the result of the ZFEL.

15. In the next chapter we will show how diffusing means result in increasing variances. That is, although the first moment of the distribution, the mean, is not *driven,* the second moment, the variance, is. In the present context that translates into nondriven movement in morphospace resulting in a driven movement in complexity space.

16. To produce a result like that in figure 5.2D, a passive mechanism, it would have to be the case that in about half of the yards the leaves tend to disperse, while in the other half they spontaneously tend to collect themselves into

a pile. In the extreme case, in some yards, the leaves would pile themselves into a single stack, representing the minimum possible dispersal in space. In such an unlikely scenario, the average and maximum dispersal of leaves across all yards would increase, but the minimum would not.

17. We do not need to assume that choanoflagellates are identical to the animal ancestor, or even that they are the survivors of an unbroken chain of persistent simple forms. For our purposes, it would not matter if they had been secondarily simplified from complex animals. The point is that they suggest that simple forms, however derived, may have existed continuously over the history of the animals. The incompleteness of the fossil record leaves open the possibility that the current occupation of the lowest level is unusual, that for most of the history of animals, single-cell-type organisms have been absent. Still, the minimum probably cannot have been much higher than about ten, the number of cell types in sponges, which have persisted continuously.

18. It would be nice to have better evidence for a trend mechanism at the largest scale, something more direct and telling than the trajectory of the minimum. Studies have been conducted in various lower taxa on trend mechanisms for complexity by sampling ancestor-descendant species pairs from the taxon and measuring the change in complexity in each pair. Using this approach, Saunders, Work, and Nikolaeva (1999) discovered a driven increasing trend in the complexity of ammonoid sutures over the 140-million-year history of the group. And Adamowicz, Purvis, and Wills (2008) found a driven increasing trend in the complexity of the arthropod limb series over its 500-million-year history. On the other hand, Sidor (2001) found a driven *decreasing* trend in the complexity of skulls in mammal-like reptiles. And McShea (1993, 1996) found no tendency toward either increase or decrease in the evolution of the mammalian vertebral column. More studies of trends in more groups—including protists, fungi, and plants, as well as animals—are necessary to know whether any particular mechanism predominates across all groups.

19. Saunders and Ho (1976) focused on the structural aspect of this story, the contribution of the structure of development to this bias in favor of complexity. But the argument clearly has a selective component as well.

CHAPTER SIX

1. This is a case of what Brandon (2005) labels a *maximal probability difference* and so one in which drift is not possible.

2. That condition is sufficient but not necessary. Many weaker conditions would suffice as well. For instance, the points could move in concert but with the rightmost points moving more rightward with each step in time. Here the variance would constantly increase.

3. We can approach the problem of generalizing the theory of drift even more abstractly. Drift requires probabilistic sampling. Thus, we can say that whenever probabilistic sampling occurs, drift *is possible*. One caveat is necessary here. Suppose we have a collection of entities to be sampled, each with a given probability of being sampled. There then is a distribution of those probabilities. The number of possible such distributions is a combinatorial function

of the number of entities in the collective. A very small subset of those distributions is what are called distributions of maximal probability difference (MPD) (Brandon 2005). In such distributions all probabilities are either 1 or 0, with at least some of each value. In MPD distributions drift is impossible. Why? Recall our definition of drift. Drift just is some outcome that differs from the probabilistic expectation. But with all of the probabilities being either 1 or 0, such a deviation from expectation is impossible. For a given collective one can arrange all possible probability distributions of the entities of the collective on a continuum with the MPD distributions being one end of the continuum and the equiprobable distribution being the other. (The equiprobable distribution assigns each entity the same probability of being sampled. For finite N, that probability would be $1/N$. Thus, for a given collective, there is only one such distribution; see Brandon 2005 for further discussion.) This way of thinking about drift is helpful in that it shows the regions of state space where drift is possible and where it is not. As can be seen, and we think easily understood given this abstract understanding of drift, drift is always possible except in the MPD zone. Metaphysical determinists will say that, although only a small part of your graph, that is where life is. Biologists, who think drift is real, will think that the metaphysician should get out more. Interesting as this is, we will pursue it no further.

4. This statement should not be confused with the similar-sounding statement that all selection is directional, which is contradicted by stabilizing and disrupting selection.

5. One might object that sophisticated evolutionary geneticists would not use the H-W law to provide the null hypothesis in more complex genetic situations (which are, of course, the norm). Rather, they would use some specific evolutionary genetic model tailored to the specific situation to provide a better, more accurate, null hypothesis. Crucially, some such models would not predict genetic stasis as the null from every point in state space. That is, if the genetic system is thought of as providing the zero-force condition, and things like selection, migration, etc. are thought of as external forces, then not all such models reach equilibrium in a single generation. For instance, in a multilocus model far from linkage equilibrium, the genetic system left to itself could take many generations to reach equilibrium. Perhaps one could devise some genetic models that produce infinite cycles without ever reaching equilibrium. Nonetheless, we do not feel that we are attacking a straw position in pointing out the problems with taking the H-W law as somehow fundamental in evolutionary theory. The citations given above justify our stance. In chapter 7, we will provide further justification for the view that stasis has been seen as the default condition of evolutionary systems.

6. It would be natural to interpret Newton's choice of inertial motion as the "default" sort of motion, or as the "zero-net-force" condition, as an objective stance, one that was vindicated by Einstein's special theory of relativity. However, many would see Einstein's general theory of relativity as refuting that position, as the principle of equivalence says that there is nothing special about inertial reference frames. It is beyond our expertise to comment on this matter except to say that it matters not for our position.

7. The sense of heritability relevant to evolutionary studies is that due to Galton ([1869] 1881). It is entirely statistical and phenomenological and makes no reference to any particular material basis. So, strictly speaking, it is not a genetic notion but does correspond to what geneticists call narrow-sense heritability (h^2), which in some, but by no means all, situations equals the fraction of the total variance that is additive genetic variance. See Roughgarden 1979 for a discussion of the conditions under which that equation holds. It is worth stating here that it is a serious error to think that evolutionary heritability just equals the additive variance divided by the total variance. That would be a far too restrictive notion.

8. Imagine a really low fidelity reproductive system, that is, one that by some absolute standard does not produce offspring that much resemble their parents. But remember that the evolutionarily relevant notion of heritability is a statistical one: it measures how much offspring resemble their parents as opposed to the population mean. Now, for any level of absolute reproductive fidelity, if one increases the level of population variation, one automatically increases the level of heritability. But, of course, we have argued that there is a built-in tendency for population variation to increase. So, heritability is to be expected. This, in short, is the rationale for our choice of characterization of evolutionary systems as ones in which there is reproduction of heritable variation.

9. G is very hard to measure accurately; see Schwarzschild 2000.

10. See Brandon 1990 for further discussion.

11. We must make clear that what we have said concerns only the statistical-mechanical interpretation of the second law. There is another second-law school in biology that invokes the energetic interpretation, focusing on somewhat-different aspects of biology, especially energy flows (Wicken 1987; Salthe 1993; Chaisson 2001). Many of the predictions of that school are similar to those of the zero-force law, including increasing diversity and complexity, but the theoretical foundations appear to be very different. At this point, the degree of overlap and of consistency with our view is unclear.

12. One of us (RB) believes that the foundations of thermodynamics are not well understood. In the philosophy of physics, the project has long been to explain how a probabilistic law (the second law) could arise out of deterministic underpinnings (the deterministic behavior of gas molecules in a container). This project, though fascinating, is ill conceived in at least two ways. First, it assumes that Newtonian mechanics is always deterministic. The truth of that, it turns out, is greatly exaggerated. Second, it assumes that Newtonian mechanics is the right reductive foundation for statistical mechanics. But surely statistical quantum mechanics is the true foundation. For these reasons, we expect no help from thermodynamics.

13. We are assuming here what we take to be the interpretation of the statistical-mechanical version of the second law that is standard among physicists. This interpretation is not hierarchical, mainly because a hierarchical perspective adds nothing to the standard atomic/molecular interpretation. Why? Because higher-level order, say at the level of crystals, is perfectly recognizable at the molecular level. That is, a lower entropic arrangement at the crystal level is also a lower entropic arrangement of the molecules from which the crystals

coalesced. Thus, the molecular perspective is more general than the hierarchical because in a pure liquid state of the relevant molecules there will be no crystals and thus no crystal-level measure of entropy; and it always gives the right answer. Perhaps some previous theorists who have tried to apply the second law to evolution have not recognized this.

14. See Spencer 1900.

CHAPTER SEVEN

1. That is, geographically isolated populations will tend to diverge without any input from selection *for* reproductive isolation. On the other hand, when allopatric speciation is produced by different selection pressures acting to some degree independently in the different populations, the resulting speciation is attributable to the ZFEL.

2. Genic selectionism is the view put forward originally by Williams (1966) and later championed by Dawkins (1976). It holds that all selection can be adequately modeled in terms of selection coefficients that attach to alternative alleles. The alternative view is hierarchical, which implies that sometimes to get a predictive and explanatory model one must assign fitness values to higher-level entities (e.g., genotypes).

3. As a consequence, some horrible misconceptions about issues concerning the levels of selection (Brandon and Nijhout 2006) have arisen.

4. There are good reasons to be unhappy with the machine metaphor for organisms, but we will ignore those for present purposes.

5. If so, the question arises why genomes are not equally streamlined. Or to put it another way, one might ask rhetorically: if organisms have many pseudo-genes, why could they not also have many nonfunctional parts at higher levels (cells, tissues, organs, etc.)?

6. We have in mind a process similar to that envisioned by Kauffman (1993) in the emergence of autocatalytic loops in sufficiently large and diverse populations of macromolecules.

7. For example, it has been proposed that the flagellar motor could have arisen from a much simpler device that bacteria use to inject toxin into cells that they are attacking, a device whose part types are a subset of those of the flagellar motor.

8. The ZFEL for complexity is about parts, not interactions. But a pure (function-free) notion of complexity could also be devised for interactions (see McShea 1996; see also chapter 4, n. 3, above), and a corresponding ZFEL could be developed. In other words, a ZFEL analog could be developed that predicts increasing complexity of interactions, for example, behaviors (Pringle 1951). In that case, the interactional aspect of "order" might be dissociable into a complexity-of-interaction piece, explained by the ZFEL analog, plus a regular and functional piece explained by a combination of self-organization and selection.

9. The precise source of the famous quotation from William James at the head of this section is hard to trace. But see the discussion in a footnote in Dennett (1995, 393), where a possible source is given as James's "Lecture Notes

1880–1897." The ZFEL has thick roots in Spencer's view, and although James intended the remark as parody, "somethingelsification" is a fair one-word descriptor of the process at the heart of the ZFEL. The original remark that James was parodying appears in Spencer's *First Principles* (1862, 216): "Evolution is a change from an indefinite, incoherent, homogeneity to a definite, coherent, heterogeneity, through continuous differentiations and integrations."

10. Recall our discussion in chapter 6. Although the probability calculus may be analytic, the law of likelihood, or the principle of direct inference, is not. The more probable does not have to happen more often than the less probable. That is a corollary of Hume's critique of induction.

11. The paleobiology literature, in particular, is replete with studies using morphospaces developed ad hoc for a phylogenetically restricted set of taxa and based on a limited number of shared characters. See, for example, a skeleton space devised by Thomas and Reif (1993). Niklas (1992) has devised such a limited morphospace for land plants.

12. Such a model might be constructed along the lines of those in Gavrilets 1999 or Pie and Weitz 2005, but with the number of taxa fixed at two.

13. Our discussion here technically takes us beyond the ZFEL, because human institutions and culture do not replicate in the same way that organisms do. In effect, we are invoking a principle analogous to the ZFEL and allied to the general version of the ZFEL, the G-ZFEL, mentioned in chapter 6.

14. There is a small evolutionary literature on quantifying the effect of drift in lineages on macroevolutionary scales (Lande 1976; Lynch 1990). The suggestion here is that this work could be extended hierarchically to encompass the equivalent of drift at the cell, organ, organism, and even clade levels.

References

Adami, C. 2002. What is complexity? *Bioessays* 24:1085–94.

Adamowicz, S. J., A. Purvis, and M. A. Wills. 2008. Increasing morphological complexity in multiple parallel lineages of the Crustacea. *Proceedings of the National Academy of Sciences* 105:4786–91.

Alroy, J. 2000. Understanding the dynamics of trends within evolving lineages. *Paleobiology* 26:319–29.

Alroy, J., et al. 2001. Effects of sampling standardization on estimates of Phanerozoic marine diversification. *Proceedings of the National Academy of Sciences* 98:6261–66.

Alroy, J., et al. 2008. Phanerozoic trends in the global diversity of marine invertebrates. *Science* 321:97–100.

Antonovics, J. N., C. Ellstrand, and R. N. Brandon. 1988. Genetic variation and environmental variation: Expectations and experiments. In *Plant Evolutionary Biology,* ed. L. D. Gottlieb and S. K. Jain, 275–303. London: Chapman & Hall.

Arthur, W. 1988. *A Theory of the Evolution of Development.* New York: Wiley.

Bailey, D. W. 1977. Genetic drift: The problem and its possible solution by frozen-embryo storage. *Ciba Foundation Symposium* 52:291–303.

Bambach, R. K. 1977. Species richness in marine benthic habitats through the Phanerozoic. *Paleobiology* 3:152–67.

———. 1983. Ecospace utilization and guilds in marine communities through the Phanerozoic. In *Biotic Interactions in Recent and Fossil Benthic Communities,* ed. M. J. S. Tevesz and P. L. McCall, 719–46. New York: Plenum Press.

———. 1985. Classes and adaptive variety: The ecology of diversification in marine faunas through the Phanerozoic. In *Phanerozoic Diversity Patterns: Profiles in Macroevolution,* ed. J. W. Valentine, 191–253. Princeton, NJ: Princeton University Press.

———. 1993. Seafood through time: Changes in biomass, energetics, and productivity in the marine ecosystem. *Paleobiology* 19:372–97.

Bambach, R. K., A. M. Bush, and D. H. Erwin. 2007. Autecology and the filling of ecospace: Key metazoan radiations. *Palaeontology* 50:1–22.

Bamshad, M., and S. P. Wooding. 2003. Signatures of natural selection in the human genome. *Nature Reviews: Genetics* 4:99–111.

Beatty, J. 1981. What's wrong with the received view of evolutionary theory? In *Philosophy of Science Association 1980,* ed. P. Asquith and R. Giere, 397–426. East Lansing, MI: Philosophy of Science Association.

Bell, G., and A. O. Mooers. 1997. Size and complexity among multicellular organisms. *Biological Journal of the Linnean Society* 60:345–63.

Bernstein, H., H. Byerly, F. Hopf, and R. E. Michod. 1985. DNA damage, mutation and the evolution of sex. *Science* 229:1277–81.

Bernstein, H., G. S. Byers, and R. E. Michod. 1981. The evolution of sexual reproduction: The importance of DNA repair, complementation, and variation. *American Naturalist* 117:537–49.

Bonner, J. T. 1988. *The Evolution of Complexity by Means of Natural Selection.* Princeton, NJ: Princeton University Press.

———. 2004. The size-complexity rule. *Evolution* 58:1883–90.

Bouchard, F. 2008. Causal processes, fitness and the differential persistence of lineages. *Philosophy of Science* 75:560–70.

Brace, C. L. 1963. Structural reduction in evolution. *American Naturalist* 97:39–49.

Brandon, R. N. 1978. Adaptation and evolutionary theory. *Studies in History and Philosophy of Science* 9:181–206.

———. 1982. The levels of selection. In *Philosophy of Science Association 1982,* ed. P. Asquith and T. Nickles, 315–23. East Lansing, MI: Philosophy of Science Association.

———. 1990. *Adaptation and Environment.* Princeton, NJ: Princeton University Press.

———. 2005. The difference between selection and drift: A reply to Millstein. *Biology and Philosophy* 20:153–70.

———. 2006. The principle of drift: Biology's first law. *Journal of Philosophy* 103:319–35.

Brandon, R. N., and S. Carson. 1996. The indeterministic character of evolutionary theory: No "no hidden variables proof" but no room for determinism either. *Philosophy of Science* 63:315–37.

Brandon, R. N., and H. F. Nijhout. 2006. The empirical nonequivalence of genic and genotype models of selection: A (decisive) refutation of genic selectionism and pluralistic genic selectionism. *Philosophy of Science* 73:277–97.

Brandon, R. N., and G. Ramsey. 2007. What's wrong with the emergentist statistical interpretation of natural selection and random drift. In *The Cambridge*

Companion to Philosophy of Biology, ed. M. Ruse and D. L. Hull, 66–84. New York: Cambridge University Press.

Brandon, R. N., and A. Rosenberg. 2003. Philosophy of biology. In *Philosophy of Science Today,* ed. P. Clark and K. Hawley, 147–80. New York: Oxford University Press,

Briggs, D. E. G., and R. A. Fortey. 2005. Wonderful strife: Systematics, stem groups, and the phylogenetic signal of the Cambrian radiation. *Paleobiology* 31 (no. 2, Supplement): 94–112.

Briggs, D. E. G., R. A. Fortey, and M. A. Wills. 1992. Morphological disparity in the Cambrian. *Science* 256:1670–73.

Brooks, D. R., and D. A. McLennan. 1993. Macroevolutionary patterns of morphological diversification among parasitic flatworms (Platyhelminthes: Cercomaria). *Evolution* 47:495–509.

Brooks, D. R., and E. O. Wiley. 1988. *Evolution as Entropy.* 2nd ed. Chicago: University of Chicago Press.

Buchholtz, E. A., and E. H. Wolkovich. 2005. Vertebral osteology and complexity in *Lagenorhynchus acutus. Marine Mammal Science* 21:411–28.

Bush, A. M., R. K. Bambach, and G. M. Daley. 2007. Changes in theoretical ecospace utilization in marine fossil assemblages between the mid-Paleozoic and late Cenozoic. *Paleobiology* 33:76–97.

Callebaut W., G. B. Müller, and S. A. Newman. 2007. The organismic systems approach: Evo-devo and the streamlining of the naturalistic agenda. In *Integrating Evolution and Development: From Theory to Practice,* ed. R. Sansom and R. N. Brandon, 25–92. Cambridge, MA: MIT Press.

Callebaut W., and D. Rasskin-Gutman, eds. 2005. *Modularity: Understanding the Development and Evolution of Natural Complex Systems.* Cambridge, MA: MIT Press.

Carroll, S. B. 2001. Chance and necessity: The evolution of morphological complexity and diversity. *Nature* 409:1102–9.

———. 2005. *Endless Forms Most Beautiful: The New Science of Evo Devo and the Making of the Animal Kingdom.* New York: W. W. Norton.

Chaisson, E. 2001. *Cosmic Evolution: The Rise of Complexity in Nature.* Cambridge, MA: Harvard University Press.

Ciampaglio, C. N., M. Kemp, and D. W. McShea. 2001. Detecting changes in morphospace occupation patterns in the fossil record: Characterization and analysis of measures of disparity. *Paleobiology* 27:695–715.

Cisne, J. L. 1974. Evolution of the world fauna of aquatic free-living arthropods. *Evolution* 28:337–66.

Collier, J. 1986. Entropy in evolution. *Biology and Philosophy* 1:5–24.

———. 2003. Hierarchical dynamical information systems with a focus on biology. *Entropy* 5:100–124.

Conway Morris, S. 1998. *The Crucible of Creation: The Burgess Shale and the Rise of Animals.* Oxford: Oxford University Press.

———. 2003. *Life's Solution: Inevitable Humans in a Lonely Universe.* Cambridge: Cambridge University Press.

Cope, E. D. 1871. The method of creation of organic forms. *Proceedings of the Amererican Philosophical Society* 12:229–63.

Crutchfield, J. P. 1992. Knowledge and meaning . . . chaos and complexity. In *Modeling Complex Phenomena,* ed. L. Lam and V. Naroditskym, 66–101. Berlin: Springer.

Crutchfield, J. P., and K. Young. 1989. Inferring statistical complexity. *Physical Review Letters* 63:105–8.

Culver, D. C., T. C. Kane, and D. W. Fong. 1995. *Adaptation and Natural Selection in Caves: The Evolution of* Gammarus minus. Cambridge, MA: Harvard University Press.

Damuth, J. 1985. Selection among "species": A formulation in terms of natural functional units. *Evolution* 39:1132–46.

Darwin, C. [1859] 1964. *On the Origin of Species (A Facsimile of the First Edition).* Cambridge, MA: Harvard University Press.

———. 1987. Notebook E. In *Charles Darwin's Notebooks,* ed. P. H. Barrett et al. Ithaca, NY: Cornell University Press.

Dawkins, R. 1976. *The Selfish Gene.* New York: Oxford University Press.

———. 1986. *The Blind Watchmaker.* New York: Norton.

Dennett, D. C. 1995. *Darwin's Dangerous Idea: Evolution and the Meanings of Life.* New York: Simon & Schuster.

Droser, M. L., and D. J. Bottjer. 1989. Ordovician increase in extent and depth of bioturbation: Implications for understanding early Paleozoic ecospace utilization. *Geology* 17:850–52.

Eble, G. J. 1999. On the dual nature of chance in evolutionary biology and paleobiology. *Paleobiology* 25:75–87.

Ehling, U. H. 1965. The frequency of X-ray induced dominant mutations affecting the skeletons of mice. *Genetics* 51:723–32.

———. 1966. Dominant mutations affecting the skeletons in offspring of X-irradiated male mice. *Genetics* 54:1381–89.

Ehrlich, M., and R. Y. Wang. 1981. 5-Methylcytosine in eukaryotic DNA. *Science* 212:1350–57.

Eldredge, N., and S. J. Gould. 1972. Punctuated equilibria: An alternative to phyletic gradualism. In *Models in Palaeobiology,* ed. T. J. M. Schopf, 82–115. San Francisco: Freeman Cooper.

Eldredge, N., and S. N. Salthe. 1984. Hierarchy and evolution. *Oxford Surveys in Evolutionary Biology* 1:182–206.

Endler, J. A. 1986. *Natural Selection in the Wild.* Princeton, NJ: Princeton University Press.

Erwin, D. H. 2001. Lessons from the past: Biotic recoveries from mass extinctions. *Proceedings of the National Academy of Sciences* 98:5399–5403.

———. 2006. *Extinction: How Life on Earth Nearly Ended 250 Million Years Ago.* Princeton, NJ: Princeton University Press.

———. 2007. Disparity: Morphological pattern and developmental context. *Palaeontology* 50:57–73.

Fisher, R. A. 1930. *The Genetical Theory of Natural Selection.* Oxford: Oxford University Press.

Fong, D. W., T. C. Kane, and D. C. Culver. 1995. Vestigialization and loss of nonfunctional characters. *Annual Review of Ecology and Systematics* 26:249–68.

Foote, M. 1996. Models of morphological diversification. In *Evolutionary Paleobiology*, ed. D. Jablonski, D. H. Erwin, and J. H. Lipps, 62–86. Chicago: University of Chicago Press.

———. 1997. The evolution of morphological diversity. *Annual Review of Ecology and Systematics* 28:129–52.

Freeling, M., and B. C. Thomas. 2006. Gene-balanced duplications, like tetraploidy, provide predictable drive to increase morphological complexity. *Genome Research* 16:805–14.

Galton, F. [1869] 1881. *Hereditary Genius: An Inquiry into Its Laws and Consequences*. New York: D. Appleton and Co.

Garcia-Fernàndez, J. 2005. Hox, parahox, protohox: Facts and guesses. *Heredity* 94:145–52.

Gavrilets, S. 1999. Dynamics of clade diversification on the morphological hypercube. *Proceedings of the Royal Society of London B* 266:817–24.

Gell-Mann, M. 1994. *The Quark and the Jaguar*. New York: Freeman.

Glymour, B. 2001. Selection, indeterminism, and evolutionary theory. *Philosophy of Science* 68:518–35.

Goodwin, B. C. 1994. *How the Leopard Changed Its Spots: The Evolution of Complexity*. New York: Charles Scribner's Sons.

Gould, S. J. 1977. *Ontogeny and Phylogeny*. Cambridge, MA: Belknap Press of Harvard University Press.

———. 1988. Trends as changes in variance: A new slant on progress and directionality in evolution. *Journal of Paleontology* 62:319–29.

———. 1989. *Wonderful Life*. New York: W. W. Norton.

———. 1991. The disparity of the Burgess Shale arthropod fauna and the limits of cladistic analysis: Why we must strive to quantify morphospace. *Paleobiology* 17:411–23.

———. 1996. *Full House: The Spread of Excellence from Plato to Darwin*. New York: Harmony Books.

———. 2002. *The Structure of Evolutionary Theory*. Cambridge, MA: Belknap Press of Harvard University Press.

Gould, S. J., and N. Eldredge. 1993. Punctuated equilibrium comes of age. *Nature* 366:223–27.

Gould, S. J., and R. C. Lewontin. 1979. The spandrels of San Marco and the Panglossian paradigm: A critique of the adaptationist programme. *Proceedings of the Royal Society of London B* 205:581–98.

Gregory, W. K. 1934. Polyisomerism and anisomerism in cranial and dental evolution among vertebrates. *Proceedings of the National Academy of Sciences* 20:1–9.

———. 1935. Reduplication in evolution. *Quarterly Review of Biology* 10:272–90.

Griesemer, J. R. 2000. Reproduction and the reduction of genetics. In *The Concept of the Gene in Development and Evolution: Historical and Epistemological Perspectives*, ed. R. F. P. Beurton and H. J. Rheinberger, 240–85. Cambridge: Cambridge University Press.

Hacking, I. 1990. *The Taming of Chance*. Cambridge: Cambridge University Press.

Hardy, G. H. 1908. Letter to the editor. *Science* 28:49–50.

Hedrick, P. W. 1986. Genetic polymorphism in heterogeneous environments: A decade later. *Annual Review of Ecology and Systematics* 17:535–66.

———. 2006. Genetic polymorphism in heterogeneous environments: The age of genomics. *Annual Review of Ecology, Evolution, and Systematics* 37:67–93.

Hedrick, P. W., M. E. Ginevan, and E. P. Ewing. 1976. Genetic polymorphism in heterogeneous environments. *Annual Review of Ecology and Systematics* 7:1–32.

Hull, D. L. 1981. Units of evolution: A metaphysical essay. In *The Philosophy of Evolution,* ed. U. L. Jensen and R. Harré, 23–44. Brighton, UK: Harvester Press.

Jablonski, D. 2005. Mass extinctions and macroevolution. *Paleobiology* 31 (no. 2, Supplement): 192–210.

———. 2008. Species selection: Theory and data. *Annual Review of Ecology and Systematics* 39:501–24.

Jacob, F. 1977. Evolution and tinkering. *Science* 196:1161–66.

Jernvall, J., and J. P. Hunter. 1995. The hypocone as a key innovation in mammalian evolution. *Proceedings of the National Academy of Sciences* 92: 10718–22.

Kanthaswamy, S., and D. G. Smith. 2002. Assessment of genetic management at three captive specific-pathogen-free rhesus macaque (*Macaca mulatta*) colonies. *Comparative Medicine* 52:414–23.

Kauffman, S. A. 1993. *The Origins of Order.* New York: Oxford University Press.

———. 2000. *Investigations.* New York: Oxford University Press.

Kimura, M. 1968. Evolutionary rate at the molecular level. *Nature* 217: 624–26.

———. 1983. *The Neutral Theory of Molecular Evolution.* Cambridge: Cambridge University Press.

Kingsolver, J. G., H. E. Hoekstra, J. M. Hoekstra, D. Berrigan, S. N. Vignieri, C. E. Hill, A. Hoang, P. Gibert, and P. Beerli. 2001. The strength of phenotypic selection in natural populations. *American Naturalist* 157:245–61.

Knoll, A. H. 2003. *Life on a Young Planet: The First Three Billion Years of Evolution on Earth.* Princeton, NJ: Princeton University Press.

Knoll, A. H., and R. K. Bambach. 2000. Directionality in the history of life: Diffusion from the left wall or repeated scaling of the right? *Paleobiology* 26:1–14.

Knoll, A. H., and S. B. Carroll. 1999. Early animal evolution: Emerging views from comparative biology and geology. *Science* 284:2129–37.

Kreitman, M. 1991. Detecting selection at the level of DNA. In *Evolution at the Molecular Level,* ed. R. K. Selander, A. G. Clark, and T. S. Whittam, 204–21. Sunderland, MA: Sinauer Associates.

———. 2000. Methods to detect selection in populations with applications to the human. *Annual Review of Genomics and Human Genetics* 1:539–59.

Kreitman, M., and R. R. Hudson. 1991. Inferring the evolutionary histories of the *Adh* and *Adh-dup* loci in *Drosophila melanogaster* from patterns of polymorphism and divergence. *Genetics* 127:565–82.

Lande, R. 1976. Natural selection and random genetic drift in phenotypic evolution. *Evolution* 30:314–34.

Leamy, L. J., and C. P. Klingenberg. 2005. The genetics and evolution of fluctuating asymmetry. *Annual Review of Ecology, Evolution, and Systematics* 36:1–21.

Lenski, R. E., C. Ofria, R. T. Pennock, and C. Adami. 2003. The evolutionary origin of complex features. *Nature* 423:139–44.

Lewontin, R. C. 1970. The units of selection. *Annual Review of Ecology and Systematics* 1:1–18.

Lloyd, E. A. 2001. Units and levels of selection. In *Thinking about Evolution: Historical, Philosophical, and Political Perspectives,* ed. R. S. Singh, C. B. Krimbas, D. B. Paul, and J. Beatty, 267–91. New York: Cambridge University Press.

Lohaus, R., N. L. Geard, J. Wiles, and R. B. R. Azevedo. 2007. A generative bias towards average complexity in artificial cell lineages. *Proceedings of the Royal Society of London B* 274:1741–50.

Lynch, M. 1990. The rate of morphological evolution in mammals from the standpoint of the neutral expectation. *American Naturalist* 136:727–41.

———. 2007a. The frailty of adaptive hypotheses for the origins of organismal complexity. *Proceedings of the National Academy of Sciences* 104:8597–8604.

———. 2007b. *The Origins of Genome Architecture*. Sunderland, MA: Sinauer Associates.

Lynch, M., and J. S. Conery. 2000. The evolutionary fate and consequences of duplicate genes. *Science* 290:1151–55.

Lynch, M. L., and B. Walsh. 1998. *Genetics and Analysis of Quantitative Traits*. Sunderland, MA: Sinauer Associates.

Marcot, J., and D. W. McShea. 2007. Increasing hierarchical complexity throughout the history of life: Phylogenetic tests of trend mechanisms. *Paleobiology* 33:182–200.

Marcus, J. M. 2005. A partial solution to the C-value paradox. *Lecture Notes in Computer Science* 3678:97–105.

Matthen, M., and A. Ariew. 2002. Two ways of thinking about fitness and natural selection. *Journal of Philosophy* 99:55–83.

Maynard Smith, J., and E. Szathmáry. 1995. *The Major Transitions in Evolution*. Oxford: Freeman.

Mayr, E. 1963. *Animal Species and Evolution*. Cambridge, MA: Harvard University Press.

McCarthy, M. C., and B. J. Enquist. 2005. Organismal size, metabolism and the evolution of complexity in metazoans. *Evolutionary Ecology Research* 7:681–96.

McDonald, J. H., and M. Kreitman. 1991. Adaptive protein evolution at the *Adh* locus in *Drosophila*. *Nature* 351:652–54.

McFadden, J. 2001. *Quantum Evolution*. New York: W. W. Norton.

McShea, D. W. 1991. Complexity and evolution: What everybody knows. *Biology and Philosophy* 6:303–24.

———. 1992. A metric for the study of evolutionary trends in the complexity of serial structures. *Biological Journal of the Linnean Society* 45:39–55.

———. 1993. Evolutionary change in the morphological complexity of the mammalian vertebral column. *Evolution* 47:730–40.

———. 1996. Metazoan complexity and evolution: Is there a trend? *Evolution* 50:477–92.

———. 2000. Functional complexity in organisms: Parts as proxies. *Biology and Philosophy* 15:641–68.

———. 2001. The hierarchical structure of organisms: A scale and documentation of a trend in the maximum. *Paleobiology* 27:405–23.

———. 2002. A complexity drain on cells in the evolution of multicellularity. *Evolution* 56:441–52.

———. 2005a. The evolution of complexity without natural selection, a possible large-scale trend of the fourth kind. *Paleobiology* 31 (no. 2, Supplement): 146–56.

———. 2005b. A universal generative tendency toward increased organismal complexity. In *Variation: A Central Concept in Biology*, ed. B. Hallgrímsson and B. Hall, 435–53. New York: Academic Press.

McShea, D. W., and E. P. Venit. 2001. What is a part? In *The Character Concept in Evolutionary Biology*, ed. G. P. Wagner, 259–84. San Diego, CA: Academic Press.

Miller, A. I. 1997. Dissecting global diversity patterns: Examples from the Ordovician radiation. *Annual Review of Ecology and Systematics* 28:85–104.

———. 2004. The Ordovician radiation: Toward a new global synthesis. In *The Great Ordovician Biodiversification Event,* ed. B. D. Webby, F. Paris, M. L. Droser, and I. G. Percival, 380–88. New York: Columbia University Press.

Miller, A. I., M. Aberhan, D. P. Buick, K. V. Bulinski, C. A. Ferguson, A. J. W. Hendy, and W. Kiessling. In press. Phanerozoic trends in the global geographic disparity of marine biotas. *Paleobiology*.

Miller, A. I., and M. Foote. 2003. Increased longevities of post-Paleozoic marine genera after mass extinctions. *Science* 302:1030–32.

Mills, S., and J. Beatty. 1979. The propensity interpretation of fitness. *Philosophy of Science* 46:263–86.

Moczek, A. P. 2008. On the origins of novelty in development and evolution. *BioEssays* 30:432–47.

Müller, G. B., and J. Streicher. 1989. Ontogeny of the syndesmosis tibiofibularis and the evolution of the bird hindlimb: A caenogenetic feature triggers phenotypic novelty. *Anatomy and Embryology* 179:327–39.

Newman, S. A., and G. B. Müller. 2000. Epigenetic mechanisms of character origination. *Journal of Experimental Zoology* 288:304–17.

Niklas, K. J. 1992. *Plant Biomechanics: An Engineering Approach to Plant Form and Function*. Chicago: University of Chicago Press.

Noor, M. A. 1995. Speciation driven by natural selection in *Drosophila*. *Nature* 375:674–75.

Noor, M. A., and J. L. Feder. 2006. Speciation genetics: Evolving approaches. *Nature Reviews: Genetics* 7:851–61.

Novack-Gottshall, P. M. 2007. Using a theoretical ecospace to quantify the ecological diversity of Paleozoic and modern marine biotas. *Paleobiology* 33:273–94.

Nunney, L. 1989. The maintenance of sex by group selection. *Evolution* 43: 245–57.

Oakley, T. H., and A. S. Rivera. 2008. Genomics and the evolutionary origins of nervous system complexity. *Current Opinion in Genetics and Development* 18: 479–92.

Ohno, S. 1970. *Evolution by Gene Duplication.* New York: Springer-Verlag.

Orr, H. A. 2000. Adaptation and the cost of complexity. *Evolution* 54:13–20.

Palmer, A. R. 2005. Antisymmetry. In *Variation: A Central Concept in Biology,* ed. B. Hallgrímsson and B. K. Hall, 359–98. New York: Academic Press.

Peters, S. E., and M. Foote. 2001. Biodiversity in the Phanerozoic: A reinterpretation. *Paleobiology* 27:583–601.

Pfennig, D. W., A. M. Rice, and R. A. Martin. 2007. Field and experimental evidence for competition's role in phenotypic divergence. *Evolution* 61:257–71.

Pfennig, K. S., and D. W. Pfennig. 2005. Character displacement as the "best of a bad situation": Fitness trade-offs resulting from selection to minimize resource and mate competition. *Evolution* 59:2200–2208.

Pie, M. R., and J. S. Weitz. 2005. A null model of morphospace occupation. *American Naturalist* 166:E1–E13.

Pourquié, O. 2003. The segmentation clock: Converting embryonic time into spatial pattern. *Science* 301:328–30.

Pringle, J. W. S. 1951. On the parallel between learning and evolution. *Behaviour* 3:174–215.

Prout, T. 1964. Observations on structural reduction in evolution. *American Naturalist* 98:239–49.

Raup, D. M. 1977. Stochastic models in evolutionary paleontology. In *Patterns of Evolution as Illustrated by the Fossil Record: Developments in Palaeontology and Stratigraphy,* ed. A. Hallam, 59–78. Amsterdam: Elsevier.

Raup, D. M., and S. J. Gould. 1974. Stochastic simulation and evolution of morphology—towards a nomothetic paleontology. *Systematic Zoology* 23: 305–22.

Raup, D. M., S. J. Gould, T. J. M. Schopf, and D. S. Simberloff. 1973. Stochastic models of phylogeny and the evolution of diversity. *Journal of Geology* 81:525–42.

Reichenbach, H. 1938. *Experience and Prediction.* Chicago: University of Chicago Press.

———. 1949. *The Theory of Probability.* Berkeley and Los Angeles: University of California Press.

Richardson, R. C., and R. M. Burian. 1992. A defense of the propensity interpretations of fitness. *PSA,* 1992, vol. 1:349–62.

Riedl, R. 1977. A systems-analytical approach to macro-evolutionary phenomena. *Quarterly Review of Biology* 52:351–70.

Rosenberg, A., and D. W. McShea. 2007. *Philosophy of Biology: A Contemporary Introduction.* New York: Routledge.

Roughgarden, J. 1979. *Theory of Population Genetics and Evolutionary Ecology: An Introduction.* New York: Macmillan.

Ruse, M. 1973. *The Philosophy of Biology.* London: Hutchinson's University Library.

———. 1996. *Monad to Man: The Concept of Progress in Evolutionary Biology.* Cambridge, MA: Harvard University Press.

Salmon, W. C. 1971. *Statistical Explanation and Statistical Relevance.* Pittsburgh, PA: University of Pittsburgh Press.

———. 1984. *Scientific Explanation and the Causal Structure of the World.* Princeton, NJ: Princeton University Press.

Salthe, S. N. 1985. *Evolving Hierarchical Systems.* New York: Columbia University Press.

———. 1993. *Development and Evolution: Complexity and Change in Biology.* Cambridge, MA: MIT Press.

Saunders, P. T., and M. W. Ho. 1976. On the increase in complexity in evolution. *Journal of Theoretical Biology* 63:375–84.

Saunders, W. B., D. M. Work, and S. V. Nikolaeva. 1999. Evolution of complexity in Paleozoic ammonoid sutures. *Science* 286:760–63.

Schluter, D. 2009. Evidence for ecological speciation and its alternative. *Science* 323:737–41.

Schopf, T. J. M., D. M. Raup, S. J. Gould, and D. S. Simberloff. 1975. Genomic versus morphologic rates of evolution; influence of morphologic complexity. *Paleobiology* 1:63–70.

Schwarzschild, B. 2000. Our knowledge of G gets worse, then better. *Physics Today* 53:21.

Sealfon, R. A. 2008. The problem with organismal complexity. MS diss., Duke University.

Sepkoski, J. J. 1978. A kinetic model of phanerozoic taxonomic diversity I. Analysis of marine orders. *Paleobiology* 4:223–51.

———. 1988. Alpha, beta, or gamma: Where does all the diversity go? *Paleobiology* 14:221–34.

Sidor, C. A. 2001. Simplification as a trend in synapsid cranial evolution. *Evolution* 55:1419–42.

Simon, H. A. 1969. The architecture of complexity. In *The Sciences of the Artificial,* by H. A. Simon, 84–118. Cambridge, MA: MIT Press.

Simpson, E. M., C. C. Linder, E. E. Sargent, M. T. Davisson, L. E. Mobraaten, and J. J. Sharp. 1997. Genetic variation among 129 substrains and its importance for targeted mutagenesis in mice. *Nature Genetics* 16:19–27.

Simpson, G. G. 1953. *The Major Features of Evolution.* New York: Columbia University Press.

Sober, E. 1984. *The Nature of Selection: Evolutionary Theory in Philosophical Focus.* Chicago: University of Chicago Press.

Sober, E., and D. S. Wilson. 1994. A critical review of philosophical work on the units of selection problem. *Philosophy of Science* 61:534–55.

Spencer, H. 1862. *First Principles.* London: Williams and Norgate.

———. 1900. *First Principles.* 6th ed. New York: D. Appleton.

Stamos, D. N. 2001. Quantum indeterminism and evolutionary biology. *Philosophy of Science* 68:164–84.

Stanley, S. M. 1973. An explanation for Cope's rule. *Evolution* 27:1–26.

———. 1979. *Macroevolution: Pattern and Process.* San Francisco: W. H. Freeman.

Stephens, C. 2004. Selection, drift, and the "forces" of evolution. *Philosophy of Science* 71:550–70.

Sterelny, K. 1999. Bacteria at the high table. *Biology and Philosophy* 14: 459–70.

Taylor, J. S., and J. Raes. 2004. Duplication and divergence: The evolution of new genes and old ideas. *Annual Review of Genetics* 38:615–43.

Thomas, R. D. K., and W.-E. Reif. 1993. The skeleton space: A finite set of organic designs. *Evolution* 47:341–60.

Thomas, R. D. K., R. M. Shearman, and G. W. Stewart. 2000. Evolutionary exploitation of design options by the first animals with hard skeletons. *Science* 288:1239–42.

Tokarski, T. R., and G. S. Hafner. 1984. Regional morphological variations within the crayfish eye. *Cell and Tissue Research* 235:387–92.

True, J. R., and S. B. Carroll. 2002. Gene co-option in physiological and morphological evolution. *Annual Review of Cell and Developmental Biology* 18:53–80.

True, J. R., and E. S. Haag. 2001. Developmental system drift and flexibility in evolutionary trajectories. *Evolution & Development* 3:109–19.

Twain, M. 1963. Cornpone opinions. In *Mark Twain on the Damned Human Race,* 24. New York: Hill & Wang.

Valentine, J. W. 1969. Patterns of taxonomic and ecological structure on the shelf benthos during Phanerozoic time. *Palaeontology* 12:684–709.

———. 1980. Determinants of diversity in higher taxonomic categories. *Paleobiology* 6:444–50.

Valentine, J. W., A. G. Collins, and C. P. Meyer. 1994. Morphological complexity increase in metazoans. *Paleobiology* 20:131–42.

Valentine, J. W., T. C. Foin, and D. Peart. 1978. A provincial model of Phanerozoic marine diversity. *Paleobiology* 4:55–66.

Van Valen, L. 1971. Adaptive zones and the orders of mammals. *Evolution* 25:420–28.

Vermeij, G. J. 1977. The Mesozoic marine revolution: Evidence from snails, predators and grazers. *Paleobiology* 3:245–58.

———. 1987. *Evolution and Escalation: An Ecological History of Life.* Princeton, NJ: Princeton University Press.

———. 1995. Economics, volcanoes, and Phanerozoic revolutions. *Paleobiology* 21:125–52.

Waddington, C. H. 1969. Paradigm for an evolutionary process. In *Towards a Theoretical Biology,* ed. C. H. Waddington, vol. 2, 106–28. Edinburgh: Edinburgh University Press.

Wagner, G. P., and L. Altenberg. 1996. Perspective: Complex adaptations and the evolution of evolvability. *Evolution* 50:967–76.

Wagner, P. J. 1996. Contrasting the underlying patterns of active trends in morphologic evolution. *Evolution* 50:990–1017.

Walsh, D. M., T. Lewens, and A. Ariew. 2002. Trials of life: Natural selection and random drift. *Philosophy of Science* 69:429–46.

Weldon, W. F. R. 1898. Opening address, section D, zoology. *Nature* 58: 499–506.

West-Eberhard, M. J. 2003. *Developmental Plasticity and Evolution*. New York: Oxford University Press.

Wicken, J. S. 1987. *Evolution, Thermodynamics, and Information: Extending the Darwinian Program*. New York: Oxford University Press.

Wilkens, H. 2007. Regressive evolution: Ontogeny and genetics of cavefish eye rudimentation. *Biological Journal of the Linnean Society* 92:287–96.

Williams, G. C. 1966. *Adaptation and Natural Selection*. Princeton, NJ: Princeton University Press.

———. 1975. *Sex and Evolution*. Princeton, NJ: Princeton University Press.

Williston, S. 1914. *Water Reptiles of the Past and Present*. Chicago: University of Chicago Press.

Wimsatt, W. C. 1974. Complexity and organization. In *Philosophy of Science Association 1972*, ed. K. F. Schaffner and R. S. Cohen, 67–86. Dordrecht: D. Reidel.

———. 1986. Developmental constraints, generative entrenchment, and the innate-acquired distinction. In *Integrating Scientific Disciplines*, ed. W. Bechtel, 185–208. Dordrecht: Martinus Nijhoff.

Wimsatt, W. C., and J. C. Schank. 1988. Two constraints on the evolution of complex adaptation and the means of their avoidance. In *Evolutionary Progress*, ed. M. Nitecki, 231–73. Chicago: University of Chicago Press.

Woodward, J. 2003. *Making Things Happen: A Theory of Causal Explanation*. New York: Oxford University Press.

Wright S. 1948. On the roles of directed and random changes in gene frequency in the genetics of populations. *Evolution* 2:279–94.

Yang, Z., and J. P. Bielawski. 2000. Statistical methods for detecting molecular adaptation. *Trends in Ecology and Evolution* 15:496–503.

Zhang, J. 2003. Evolution by gene duplication: An update. *Trends in Ecology and Evolution* 18:292–98.

Index

Page numbers followed by f *or* t *refer to figures or tables, respectively.*